Strategy and Tactics

William R. Talbert, editor

NAVAL INSTITUTE PRESS

Annapolis, Maryland

Naval Institute Press
291 Wood Road
Annapolis, MD 21402

NS 300 : strategy and tactics / William R. Talbert, editor. – 2nd ed.
    p. cm.
  Rev. ed. of: NS 310 / Mark Mill and Alex Mamikonian, editors.
  Includes index.
  ISBN 978-1-59114-857-9 (alk. paper)
  1. Naval strategy–Textbooks. 2. Naval tactics–Textbooks.
  I. Talbert, William R. II. NS 310. III. Title: NS300. IV. Title: Strategy and tactics.
  V163.N75 2009
  359.4–dc22
                                                                    2009027769

Printed in the United States of America on acid-free paper ∞

14 13 12 11 10 09   9 8 7 6 5 4 3 2
First printing

Contents

CHAPTER 1

Naval Doctrine

LEARNING OBJECTIVES

At the end of this chapter the student will be able to:

- Describe attrition warfare and provide a detailed example.
- Describe maneuver warfare and provide a detailed example.
- Explain both center of gravity and critical vulnerability.
- Provide an example of how a critical vulnerability can be exploited.
- Explain both focus of effort and main effort.
- Describe commander's intent.
- Explain tempo and what it accomplishes for military forces.
- Describe the nine principles of war and apply them to an example.

ADDITIONAL READING

Clausewitz, Karl von. *On War*. Ed. and trans. Michael Howard and Peter Paret (Princeton, NJ.: Princeton University Press, 1976).

Corbett, Sir Julian S. *Some Principles of Maritime Strategy* (Annapolis, Md.: Naval Institute Press, 1988).

Liddell Hart, B. H. *Strategy,* 2nd revised edition (New York: Dutton, 1991).

Wylie, Joseph C. Rear Admiral, USN (Ret.) *Military Strategy: A General Theory of Power Control* (Annapolis, Md.: Naval Institute Press, 1989).

Mahan, Alfred Thayer. *Mahan On Naval Strategy*. Ed. J. B. Hattendorf (Annapolis, Md.: Naval Institute Press, 1991).

Sun Tzu. *The Art of War*. Ed. and trans. Samuel. B. Griffith (New York: Oxford University Press, 1988).

THE BATTLE OF THE ATLANTIC

Using Attrition Warfare

In World War II, allied naval forces engaged in attrition warfare by employing their resources against the German undersea fleet. Analyzing the effectiveness of submarine warfare, the former Soviet Union Admiral of the Fleet, Sergei Gorshkov, noted in his study of this period that German submarines nearly ended the war through the rapid destruction of the allied merchant fleet. German forces, especially U-boats, were credited with sinking more than 2,800 merchant ships—68% of all tonnage sunk by Nazi Germany in the war. So devastating was this weapon that, at the height of the allied counteroffensive, for each German U-boat, there were 25 U.S. and British warships and 100 aircraft in pursuit. For every German submariner at sea, there were 100 American and British anti-submariners. A total of six million men, 5,500 specially constructed ships, and 20,000 small craft were dedicated to the antisubmarine war. As the allies pressed their offensive, Germany's losses exceeded its war industry's capacity to keep pace. At the same time, the allies were able to replace their damaged merchant fleet and even expanded it by adding replacements numbering twice the losses suffered. In the Battle of the Atlantic, the threat of the U-boat was checked by overwhelming allied response. This resource-intensive, time-consuming effort was an effective use of attrition warfare.

Introduction

The last thing that an explorer arrives at is a complete map that will cover the whole ground he has traveled, but for those who come after him and would profit by and extend his knowledge, his map is the first thing with which they will begin. So it is with strategy. . . . It is for this reason that in the study of war we must get our theory clear before we can venture in search of practical conclusions.
—*Sir Julian Corbett, 1911*

War is an instrument of a nation's power, initiated to achieve national objectives when other means to resolve differences have failed. Our fundamental military purpose is to attain national policy objectives through our capacity to wage war successfully. How well we in the Naval Services accomplish our mission depends on how thoroughly we understand both the nature and the conduct of war and learn war's many lessons. Only through such understanding can we prepare ourselves for its tests.

Two Styles of Warfare

Naval forces have followed several styles or philosophies of warfare throughout history. Two specific types—attrition and maneuver—have evolved in response to particular needs and force capabilities.

Although they vary significantly in efficiency, flexibility, and decisiveness, each type of warfare has its own utility, depending on circumstances, and both types are conducted today.

ATTRITION WARFARE

A key difference between attrition warfare—the wearing down of an enemy—and maneuver warfare—a high tempo, indirect philosophy—is our method of engaging the enemy. In the days of sail, fleet "line" tactics were much less involved. Ships in single lines exchanged heavy broadsides against an enemy similarly arrayed, all within sight of each other. Their simple doctrine called for sailing directly to the enemy's location and systematically engaging his fleet. Attrition warfare is the application of our strength against an enemy's strength. It is typically a "linear" or two-dimensional style of fighting that is frequently indecisive and inherently costly in terms of personnel, resources, and time. When success in war on the operational and strategic levels depends on our ability to destroy or deny the enemy crucial resources faster than he can recover, we are employing classic attrition warfare techniques. We attrite the enemy through systematic application of overwhelming force that reduces his ability or capacity to resist.

Maneuver Warfare

Naval forces also have used the preferable and more effective—albeit more difficult to master—fighting style known as maneuver warfare. Closely associated with the writings of Sun Tzu and used by the great practitioners of expeditionary, naval, and land war, maneuver warfare is a philosophy, rather than a formula—an approach, rather than a recipe. Like attrition warfare, it has long served as common doctrine for naval forces. It emphasizes the need to give the commander freedom to deal with specific situations. Maneuver warfare is further characterized by adaptability and is not limited to a particular environment. Though enhanced by a variety of technologies, it is not dependent upon any one of them. Maneuver warfare emphasizes the indirect approach—not merely in terms of mobility and spatial movement, but also in terms of time and our ability to take action before the enemy can counter us. Maneuver warfare requires us to project combat power. Unlike attrition warfare, however, this power is focused on key enemy weaknesses and vulnerabilities that allow us to strike the source of his power—the key to his existence and strength as a military threat.

THE CONDUCT OF WAR

Hold the attention of your enemy with a minimum force, then quickly strike him suddenly and hard on his flank or rear with every weapon you have.
—*General A. A. Vandegrift, USMC* Battle Doctrine for Front Line Leaders, 1944

Success in war often is the result of decisive action that destroys the enemy's will or capacity to resist. Because protracted war can cause high casualties and unwanted political and economic consequences, the rapid conclusion of hostilities is a key goal. Maneuver warfare, based on the twin pillars of decisiveness and rapidity, is our preferred style of warfighting. It is as applicable today in the maritime environment as it has been in traditional land warfare. Modern maneuver warfare requires integration and understanding of four key concepts—center of gravity, critical vulnerability, focus of effort, and main effort. We convey these concepts in context to our forces using a mechanism called the commander's intent.

Center of Gravity and Critical Vulnerability

The center of gravity is something the enemy must have to continue military operations—a source of his strength, but not necessarily strong or a strength in itself. There can only be one center of gravity. Once identified, we focus all aspects of our military, economic, diplomatic, and political strengths against it. As an example, a lengthy resupply line supporting forces engaged at a distance from the home front could be an enemy's center of gravity. The resupply line is something the enemy must have—a source of strength—but not necessarily capable of protecting itself. Opportunities to access and destroy a center of gravity are called critical vulnerabilities. To deliver a decisive blow to the enemy's center of gravity, we must strike at objectives affecting the center of gravity that are both critical to the enemy's ability to fight and vulnerable to our offensive actions. If the object of a strike is not critical—essential to the enemy's

INCHON–SEOUL

Using Maneuver Warfare

The Navy and the Marines have never shone more brightly than this morning.
—General Douglas MacArthur, 15 September 1950

The amphibious operation at Inchon in the Korean War was a classic example of how the naval Services have employed maneuver warfare. Prior to the operation, the North Korean Peoples' Army had driven the U.S. and allied forces into a constricted corner of South Korea and threatened to push them from the peninsula altogether. Even though his forces were in dire straits, General Douglas MacArthur, Supreme Commander of the United Nations forces in Korea, recognized that the Naval Services in his command had the ability to reverse dramatically the tide of the battle. A landing on the Korean peninsula north of the enemy lines, he reasoned, would allow his forces to sever the critical north/south rail and road supply lines running through nearby Seoul that provided vital support to the North Korean siege of the Pusan perimeter. By 15 September 1950, U.S. Navy surface combatants and carrier air squadrons, along with shore-based Marine and Air Force air units, had cleared Korean waters and air space of North Korean opposition. Thus protected and concealed from the enemy, Vice Admiral Arthur D. Struble's 260-ship, Joint Task Force Seven transported Army and South Korean ground units and the amphibious-trained 1st Marine Division to the strategically important port of Inchon, north of enemy lines. These troops stormed ashore via lanes cleared of obstructions by naval underwater demolition teams and behind the gunfire of four cruisers, eight destroyers, and the aircraft of six carriers. Amphibious support ships soon brought in reinforcements and the supplies needed to maintain and expand the beachhead. This bold, surprise maneuver severed the lines of communications to 90 percent of the enemy's ground forces positioned far to the south opposite the U.N.'s Pusan perimeter. By the end of September, faced with entrapment and almost certain destruction, the North Korean Peoples' Army fled the Republic of Korea, a nation they had invaded so eagerly only a few months before.

ability to stay in the fight—the best result we can achieve is some reduction in the enemy's strength. Similarly, if the object of a strike is not vulnerable to attack by our forces, then any attempts to seize or destroy it will be futile.

The appearance of critical vulnerabilities depends entirely upon the situation and specific objective. Some—such as electrical power generation and distribution facilities ashore or the fleet oilers supporting a task group—may be obvious. On a strategic level, examples may include a nation's dependence on a certain raw material imported by sea to support its warfighting industry, or its dependence on a single source of intelligence data as the primary basis for its decisions. Alternatively, a critical vulnerability might be an intangible, such as morale. In any case, we define critical vulnerabilities by the central role they play in maintaining or supporting the enemy's center of gravity and, ultimately, his ability to resist. We should not attempt to always designate one thing or another as a critical vulnerability. A critical vulnerability frequently is transitory or time-sensitive. Some things, such as the political will to resist, may always be critical, but will be vulnerable only infrequently. Other things, such as capital cities or an opponent's fleet, may often be vulnerable, but are not always critical. What is critical will depend on the situation. What is vulnerable may change from one hour to the next. Something may be both critical and vulnerable for a brief time only. The commander's challenge is to identify quickly enemy strengths

YORKTOWN

Exploiting a Critical Vulnerability

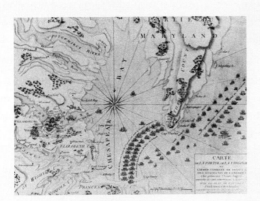

During the Revolutionary War, British forces in North America depended on free use of the adjacent seas to move and resupply their ground troops. This became especially critical to the British ability to continue fighting in August 1781, on the peninsula between Virginia's York and James Rivers, when American land forces successfully severed the British Army under General Lord Cornwallis from their ground-based resupply. At this location, British resupply by sea was vulnerable because access to the Yorktown port could be denied by controlling entry at the mouth of the Chesapeake Bay. The French West Indian Fleet under Rear Admiral François de Grasse positioned itself at this strategic location in advance of the British fleet. When British Admiral Thomas Graves arrived to support Cornwallis, de Grasse maneuvered his ships to engage the enemy outside the bay. His actions not only denied Cornwallis his needed support, but permitted another French squadron sailing from Rhode Island to enter the bay and reinforce American and French land forces. As a result, the British succumbed at Yorktown surrendering their entire Army of 7,600 men. The Franco-American alliance was effective in blocking British access to and from the sea and thereby exploiting this critical vulnerability. Losing their ability to sustain their forces by sea doomed the British war effort in North America.

and weaknesses, and recognize critical vulnerabilities when they appear. He must rapidly devise plans to avoid the strengths, exploit the weaknesses, and direct the focus of effort toward attacking the critical vulnerabilities so that he can ultimately collapse the enemy's center of gravity.

Focus of Effort and Main Effort

The focus of effort is the paramount objective to be accomplished by the force and is therefore always on the critical vulnerability that will expose the enemy's center of gravity. Since we concentrate all our resources and energy on that objective, designating the focus of effort is an important decision requiring the acceptance of risk. Responsibility for attaining the focus of effort lies with the main effort. A commander unifies the force toward the focus of effort by assigning one unit or group as the main effort.

The main effort is supported directly and indirectly by all parts of the force. When all elements of the force are focused, the strengths of each element can be brought to bear on the enemy effectively. There is only one main effort at a time and it is always directed against the focus of effort. Designating a main effort does not imply that the offensive is limited to a single attack or series of attacks. A commander may shift designation of the main effort as necessary and that designation may assign the bulk of the force or only a small fraction of the resources available. Whatever the size, designation as the main effort means that this element is central to the complete success of the operation and supporting units are obligated to do everything they can to ensure that the main effort succeeds. Supporting units are crucial to the success of mission. Leaders of supporting units, guided by the commander's intent, choose actions aimed at doing all they can to support the main effort.

Commander's Intent

Decisive action requires unity of effort—getting all parts of a force to work together. Rapid action, on the other hand, requires a large degree of decentralization, giving those closest to the problem the freedom to solve it. To reconcile these seemingly contradictory requirements, we use our understanding of the main effort and a tool called the commander's intent. The commander's intent conveys the "end state," his desired result of action. The concept of operations details the commander's estimated sequence of actions to achieve this end state and contains essential elements of a plan—i.e., what is to be done and how the commander plans to do it. A commander issues the concept of operations as part of a formal operation plan or order. The commander's intent differs from the concept of operations; a significant change in the situation that requires action often will alter the concept of operations, but the commander's intent is overarching and usually remains unchanged. The commander's intent reflects his vision and conveys his thinking through mission-type orders, in which subordinates are encouraged to exercise initiative and are given the freedom to act independently.

Mission-type orders define the contract that the commander's intent establishes between the delegating commander and his subordinates. We achieve unity of effort by promulgating the commander's intent, designating a focus of effort, and training subordinates to think in terms of the effect of their actions "two levels up" and "two levels down" in the chain of command. Since stereotyped actions are inherently predictable and thus easily countered, commanders must tailor their actions to the situation at hand, using initiative, imagination and experienced judgment.

Effective commanders at all levels neither expect nor attempt to control every action of their subordinates. Nor do they profess to foresee or attempt to plan for each contingency. Two great commanders in naval history, Admirals Horatio Nelson and Arleigh Burke, rarely issued detailed instructions to their subordinate commanders. Instead, they frequently gathered their captains to discuss a variety of tactical problems. Because of these informal discussions, the captains became aware of what their commanders expected to accomplish and how they planned, in various situations, to accomplish it. Thus prepared, they later were able to act independently, following their commanders' intent, even though formal orders either were brief or nonexistent.

The commander's intent is particularly important in cases where the situation that gave rise to orders has changed and, as a result, the original orders are no longer applicable. In such cases, subordinates can structure their decisions by asking such questions as

"What would my commander want me to do in this situation?" and "What can I do to help my commander attain the objectives?"

Tempo

Using the philosophy of maneuver warfare, we destroy or eliminate an adversary's center of gravity indirectly by attacking weaknesses or vulnerabilities that are vital to his source of power. One method of indirect attack is to create a dilemma, by putting the enemy in a situation where any step taken to counteract one threat increases his vulnerability to another. This is an indirect approach. Through rapid high-tempo actions, we present him with a series of unexpected situations and developments, each of which demands a response. In the ideal situation, the enemy would find that his best counter in one situation puts him at unacceptable risk in another—a no-win situation.

A powerful enemy can protect his critical vulnerabilities. A skillful enemy may disperse them. In each case, there is little chance of striking a decisive blow unless such an enemy can be forced to expose one or more of his critical vulnerabilities. One way of doing this is to exploit the dynamics of warfighting by maintaining a high tempo. Tempo is the pace of action—the rate at which we drive events. A rapid tempo requires that commanders be provided timely, accurate intelligence to find enemy weaknesses,

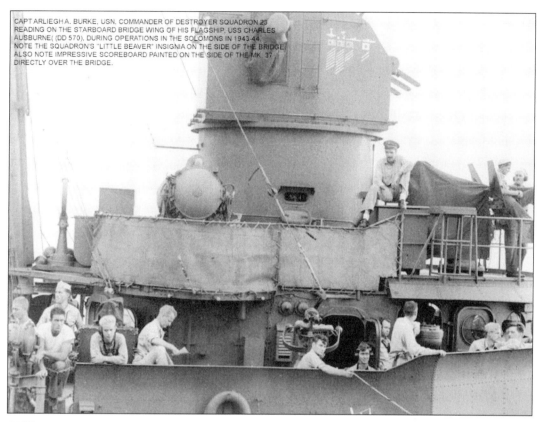

CAPT Arleigh A. Burke, Commander of Destroyer Squadron 23, reading on the bridge wing of his flagship, USS *Charles Ausburne* (DD-570), 1943.

enough decentralization to allow subordinate commanders to exploit opportunities, and clearly understood and well-rehearsed procedures at the lowest levels.

The decision cycle is a vital aspect of tempo. Forces with rapid decision cycles enjoy an advantage over those whose leaders need more time to gather and process information before making decisions. Tempo is more than a means to employ weapons better; it is a weapon itself. Directed against an enemy with a slower decision cycle, a series of rapid and unexpected attacks on critical vulnerabilities can be overwhelming, depriving him of his power to react effectively and ultimately destroying his center of gravity. As in the martial art of judo, the objective in fighting with a high tempo is to take action that sets in motion a series of actions and reactions, each of which potentially exposes—if only for a brief time—a critical vulnerability. In such a contest, we achieve victory by making the most rapid and unpredictable moves specifically selected to catch the enemy in a vulnerable position long enough to deliver a decisive blow. It is an aggressive style of warfare in which we gain advantage by observing the enemy, orienting ourselves to these surroundings, deciding on a move, and acting more rapidly than the enemy. On a tactical level, this warfighting technique, formally noted in the extraordinary success enjoyed by U.S. pilots during the Korean War, also served as the root of success in similar experiences of naval aviators during the latter stages of the Vietnam War.

Because tempo is so important in maneuver warfare, commanders must have the freedom of action to make decisions and execute them without any externally imposed delay. Commanders must be allowed to seize the initiative and respond to rapidly changing situations. Response time is a key to maneuver warfare. Activity at the operational level must contribute directly to the military strategic aim. Such aims, broadly set, demand that the operational commander have wide-ranging independence to

---

BURKE'S FIGHTING DOCTRINE

If it will help kill [The Enemy]—it's important.

If it will not help kill [The Enemy]—it's not important.

Keep your ship trained for battle!

Keep your material ready for battle!

Keep your boss informed concerning your readiness for battle!

---

exercise creativity and originality. Such freedom allows him to gain and retain the initiative and adapt to the developing situation. Mission-type orders, specifying a result but leaving open the methods of attaining that result, allow the decentralization necessary for local rapid response.

Success in war depends upon properly implementing our overall warfighting philosophy which includes understanding the commander's intent and the concepts of center of gravity, critical vulnerabilities, focus of effort, and main effort. Additionally, we must correctly apply the basic tenets or principles of war. The principles of war are based on hard-won and often bitter experience gained in conflict. These important lessons emphasize its nature and form the basis for our warfighting doctrine.

The Principles of War

An important issue throughout military history has been the way a military organization addresses the qualities that war demands from its participants. Military leadership has dealt best with the intractable problems of war as a form of military and naval art. In the maritime environment, with its distinctive factors, we fight using the principles that apply to combat everywhere. Wisdom gained from study of the basic principles of war underscores that war is not the business of managers with checklists; it is the art of leaders.

Objective. *Direct every military operation toward a clearly defined, decisive, and attainable objective.* The naval Services focus their operations to achieve political purposes defined by the National Command Authorities. With national strategic purpose identified, we can select theater military objectives and form operational and tactical objectives based on specific missions and capabilities. Whether the objective is destroying an enemy's armed forces or merely disrupting his ability to use his forces effectively, the most significant preparation a commander can make is to express clearly the objective of the operation to subordinate commanders.

Mass. *Concentrate combat power at the decisive time and place.* Use strength against weakness. A force, even one smaller than its adversary, can achieve decisive results when it concentrates or focuses its assets on defeating an enemy's critical vulnerability. A naval task force, using the sea as an ally, can compensate for numerical inferiority through the principle of mass. Mass further implies an ability to sustain momentum for decisive result.

Maneuver. *Place the enemy in a position of disadvantage through the feasible application of combat power.* Use of maneuver (mobility) capitalizes on the speed and agility of our forces (platforms and

AIR COMBAT MANEUVERING

Tactical Use of Tempo

During the early stages of the Vietnam War, our aircraft exchange rate in combat was only two to one. Air-to-air missiles, thought to be the technological answer to future aerial combat, were ineffective in many cases. Our pilots needed to develop close-in maneuvering skills and proficiency in the use of their missiles as well as newly installed guns to counter the principal communist fighters, the MiG-series.

In several traditional measures of aircraft performance the MiG was superior to the U.S. F-4. However, following the lessons taught at Top Gun—the Navy Fighter Weapons School established to study and improve air combat maneuvering skills—fighter crews improved the kill ratio sixfold in the skies over Vietnam. The F-4 crew forced its opponent into a series of tactical actions designed to gain and maintain advantage after each maneuver. The F-4 crew quickly saw how the situation changed and immediately followed with new actions. With each change, the MiG's actions became more inappropriate, until it gave the F-4 an acceptable firing opportunity. Occasionally, the MiG pilot realized what was happening to him, panicked, and ultimately made the F-4 crew's job that much easier. Success resulted from conducting a series of sudden unexpected moves to which the enemy could not adjust.

weapons) to gain an advantage in time and space relative to the enemy's vulnerabilities. Whether seen in historic warships "crossing the T," or modern ground forces enveloping an enemy, or forcing the tempo of combat beyond an adversary's ability to respond, maneuver allows us to get ahead of the enemy in several dimensions. Our advantage comes from exploiting the maneuver differential—our superiority in speed and position relative to our adversary.

Offensive. *Seize, retain, and exploit the initiative.* Since the days of sail—racing an opponent for the upwind advantage to take the initiative—offensive action has allowed us to set the terms and select the place of confrontation, exploit vulnerabilities and seize opportunities from unexpected developments. Taking the offensive through initiative is a philosophy we use to employ available forces intelligently to deny an enemy his freedom of action.

Economy of Force. *Employ all combat power available in the most effective way possible; allocate minimum essential combat power to secondary efforts.* With many more available targets than assets, each unit must focus its attention on the primary objectives. A successfully coordinated naval strike at an enemy's critical vulnerability—for example, knocking specific command-and-control nodes out of commission—can have far more significance than an attempt to destroy the entire command-and-control system.

Unity of Command. *Ensure unity of effort for every objective under one responsible commander.* Whether the scope of responsibility involves a single, independent ship at sea or the conduct of an

The Principles of War Applied at Sea

After the Battle of the Coral Sea, 4–8 May 1942, Admiral Chester W. Nimitz, Commander in Chief, U.S. Pacific Fleet, learned from signals intelligence that a large Japanese naval force, led by Admirals Isoroku Yamamoto and Chuichi Nagumo, would attack Midway, a strategic atoll west of the American fleet base at Pearl Harbor, Hawaii. Other enemy forces would make a feint toward the Aleutian Islands in the North Pacific. The priceless advantage afforded by intercepting Japanese communications gave the Americans unprecedented knowledge of enemy intentions and force dispositions.

Admiral Frank Jack Fletcher

Admiral Raymond A. Spruance

Every available carrier and escort the United States could muster was assigned to the operation—including the carrier *Yorktown,* which made a hasty sortie after repairs thought impossible by the Japanese Naval Staff. Nevertheless, the U.S. force was numerically inferior to the Japanese striking group. Nimitz assigned Rear Admiral Frank Jack Fletcher, a veteran of battle who had recently faced Japanese carrier forces at Coral Sea, as the officer in tactical command. Nimitz's ***objectives*** were clear and ***simple***: "hold Midway and inflict maximum damage on the enemy by strong attrition tactics." Nimitz further added "In carrying out the task assigned . . . you will be guided by the principles of calculated risk." Fletcher had unity of command and broad latitude in executing his tasks. He directed Rear Admiral Raymond A. Spruance, Commander Task Force 16, to attack the enemy carriers as soon as the ships were located. Fletcher, embarked in the carrier *Yorktown* with Task Force 17, would follow soon afterward. Early in the battle, when enemy air attacks placed his flagship out of action, Fletcher transferred that ***unity of command*** to Spruance who retained tactical control for most of the fight.

Knowledge of the Japanese plan allowed Nimitz to invoke ***economy of force*** by deploying minimal forces in front of a Japanese diversion toward the Aleutian Islands while ***massing*** his most effective combat power—his three aircraft carriers—against the main enemy thrust at Midway. Also, knowing that the Japanese would use submarines and long-range flying boats to determine if the U.S. fleet had sortied from Pearl Harbor, Nimitz used ***maneuver*** to frustrate the operation of these enemy units. With our intelligence advantage, the U.S. carriers were able to deploy and were in place well in advance of the enemy fleet. To retain their advantage, U.S. units maintained ***security*** through radio silence and darken-ship procedures at night. The fact that the U.S. carriers had departed the base before the battle was not known to Yamamoto. The Japanese were also conscious of the need for security and surprise. In contrast, however, excessive emphasis on security and surprise actually worked against Yamamoto and Nagumo. Convinced that the invading force would catch the island of Midway unprepared, the Japanese admirals failed to assess fully the size and location of their opposing forces.

Complete reconnaissance would have shown that the U.S. Navy did not have adequate fleet strength at the time to win in a direct at-sea confrontation. The Japanese could have concentrated their efforts against Fletcher's and Spruance's forces and then attacked the lightly defended Midway later.

On the morning of June 4, 1942, Nagumo launched a routine, limited dawn air search, convinced that the Americans could not be in the vicinity. He then followed with his initial attack against Midway, opposed only by the relatively few ground-based Navy, Marine Corps and Army Air Corps search, attack, and fighter aircraft on the island. By the time Japanese reconnaissance aircraft did discover the presence of the American force, it was too late. After the Japanese aerial assault,

Admiral Isoroku Yamamoto

Admiral Chuichi Nagumo

Spruance and his staff reasoned that Yamamoto's force might be in the process of recovering their aircraft and preparing for additional land attacks. Seizing the initiative, Fletcher and Spruance immediately attacked the Japanese carriers with every aircraft available. Although outnumbered, Fletcher and Spruance maintained an aggressive **offensive**. Japanese combat air patrol intercepted the U.S. attack, but became preoccupied with low-flying torpedo planes. When the dive bombers from *Yorktown* and *Enterprise* arrived at the battle site, the fight was taking place at low altitude, allowing them to attack Yamamoto's force unimpeded. In fact, the American air strike did surprise the Japanese carriers in an exceptionally vulnerable situation—with unstowed ordnance and bomb- and torpedo-laden planes on deck being refueled. In the fighting that followed, the Japanese lost the carriers *Hiryu, Soryu, Akagi,* and *Kaga* and their scores of veteran aviators. Deprived of air cover, Admiral Yamamoto canceled the planned invasion of Midway Island. The Japanese never regained the initiative in the Pacific.

Ships of the U.S. Pacific Fleet, 1944, Marshall Islands.

amphibious landing, we achieve unity in forces by assigning a single commander. After he expresses his intent and provides an overall focus, he permits subordinate commanders to make timely, critical decisions and maintain a high tempo in pursuit of a unified objective. The result is success, generated by unity in purpose, unit cohesion, and flexibility in responding to the uncertainties of combat.

Simplicity. *Avoid unnecessary complexity in preparing, planning, and conducting military operations.* The implementing orders for some of the most influential naval battles ever fought have been little more than a paragraph. Broad guidance rather than detailed and involved instructions promote flexibility and simplicity. Simple plans and clear direction promote understanding and minimize confusion. Operation Order 91-001, dated 17 January 1991 summarized the allied objectives for the Desert Storm campaign into a single sentence: "Attack Iraqi political-military leadership and command and control; sever Iraqi supply lines; destroy chemical, biological and nuclear capability; destroy Republican Guard forces in the Kuwaiti Theater; liberate Kuwait." These objectives were succinct, tangible, and limited.

Surprise. *Strike the enemy at a time or place or in a manner for which he is unprepared.* Catching the enemy off guard immediately puts him on the defensive, allowing us to drive events. The element of surprise is desirable, but it is not essential that the enemy be taken completely unaware—only that he becomes aware too late to react effectively. Concealing our capabilities and intentions by using covert techniques and deceptions gives us the opportunity to strike the enemy when he is not ready.

Security. *Never permit the enemy to acquire unexpected advantage. Protecting the force increases our combat power.* The alert watchstander, advanced picket, or such measures as electronic emission control all promote our freedom of action by reducing our vulnerability to hostile acts, influence, or surprise. Tools such as gaming and simulation allow us to look at ourselves from the enemy's perspective. We enhance our security by a thorough understanding of the enemy's strategy, doctrine, and tactics.

The principles of war have been proven effective in preparing for combat, but the complexities and disorder of war preclude their use as a simple checklist. Instead, we must be able to apply these principles in war's turbulent environment, to promote initiative, supplement professional judgment, and serve as the conceptual framework in which we evaluate the choices available in battle. These principles provide a solid basis for our warfighting doctrine that complements the experience and operational skill of our commanders by describing a flow of action toward objectives, rather than prescribing specific action at each point along the way. In a chaotic combat environment, doctrine has a cohesive effect on our forces, while enabling us to create disorder among our adversaries. It also promotes mutually understood terminology, relationships, responsibilities, and processes, thus freeing the commander to focus on the overall conduct of war.

REVIEW QUESTIONS

1. Describe attrition warfare and provide a detailed example.
2. Describe maneuver warfare and provide a detailed example.
3. Describe the administrative and operational Chains of Command for the Marine Corps.
4. Explain both center of gravity and critical vulnerability.
5. Provide an example of how a critical vulnerability can be exploited.
6. Describe commander's intent.
7. Explain tempo and what it accomplishes for military forces.
8. Describe the nine principles of war and apply them to an example.

CHAPTER 2

Levels of Warfare Across the Spectrum of Maritime Operations

LEARNING OBJECTIVES

The student shall be able to:

- Describe the strategic level of warfare.
- Describe the operational level of warfare.
- Describe and provide examples of the tactical level of warfare.
- Describe the primary categories within the Range of Military Operations and provide examples of each.

ADDITIONAL READING

Joint Publication 3

Introduction

The levels of war are doctrinal perspectives to clarify the links between strategic objectives and tactical actions. The three levels are strategic, operational, and tactical. Understanding the interdependent relationship of all three helps commanders visualize a logical flow of operations, allocate resources, and assign tasks. Actions within the three levels are not associated with a particular command level, unit size, equipment type, or force or component type. Instead, actions are defined as strategic, operational, or tactical based on their impact or contribution to achieving strategic, operational, or tactical objectives.

There are no finite limits or boundaries among the three levels. National assets such as intelligence and communications satellites, previously considered principally in a strategic context, today are also significant resources for tactical operations. Commanders at every level must be aware that in a world of constant, immediate communications, any single action may have consequences at all levels.

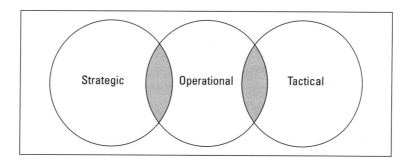

The operational level is the level of war at which campaigns and major operations are planned, conducted, and sustained to achieve strategic objectives within theaters or other operational areas (OAs). Activities at the operational level:

1. Link tactics and strategy by establishing operational objectives needed to achieve the strategic objectives.
2. Sequence events to achieve the operational objectives.
3. Initiate actions and apply resources to bring about and sustain these events.

The operational level lies between the tactical and strategic levels of war. The boundaries between the tactical, operational, and strategic levels of war overlap and are displayed as three circles with sides overlapping each other (see Figure 2-1). The overlap represents shared activities. Depending on the mission, these shared activities may be many or few.

Strategic Level of War

The strategic level is that level of war at which a nation, often as a member of a group of nations, determines national or multinational (alliance or coalition) strategic objectives and guidance and develops and uses national instruments of power to achieve these objectives. The President establishes policy, which the Secretary of State (SECSTATE) and Secretary of Defense (SecDef) translate into national strategic objectives that facilitate theater-strategic planning. Combatant commanders (CCDRs) usually participate in strategic discussions with the President and SecDef through the Chairman of the Joint Chiefs of Staff (CJCS) and with allies and coalition partners. Thus, the CCDR strategy is an element that relates to U.S. national strategy and operational activities within the theater. Derived from national strategy and policy and shaped by doctrine, military strategy provides a framework for conducting operations. Strategic military objectives define the role of military forces in the larger context of national strategic objectives.

For specific situations that require the employment of military capabilities (particularly for anticipated major combat operations), the President and SecDef typically establish a set of national strategic objectives. The supported CCDR often will have a role in achieving more than one national objective. Some national objectives will be the primary responsibility of the CCDR, while others will require a more balanced use of all instruments of national power with the CCDR in support of other government agencies. Achievement of these objectives should result in attainment of the national strategic end state—the broadly expressed conditions that should exist at the end of a campaign or operation. Once established, the national strategic objectives enable the supported commander to develop the military end state, recommended termination criteria, and supporting military strategic objectives.

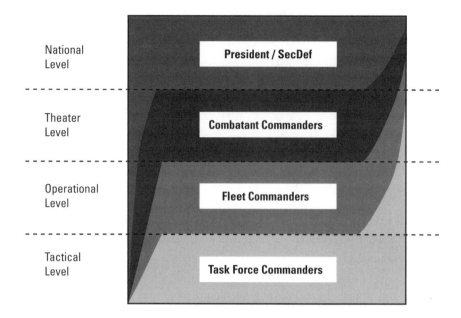

Objective 1. The clearly defined, decisive, and attainable goal toward which every operation is directed. 2. The specific target of the action taken (for example, a definite terrain feature; the seizure or holding of which is essential to the commander's plan; or an enemy force or capability without regard to terrain features). (JP 1-02. Source: JP 5-0)

Commanders at the strategic level define the military end state using military strategic objectives and the conditions that can support achievement of each objective. The strategic commander defines the time and space along with his military strategic objectives and conditions in his guidance to the operational commander.

Operational Level

The operational level links the tactical employment of forces to national and military strategic objectives. The focus at this level is on the design and conduct of operations using operational art, which is defined in JP 1-02, DOD Dictionary of Military and Associated Terms, as "the application of creative imagination by commanders and staffs—supported by their skill, knowledge, and experience—to design strategies, campaigns, and major operations and organize and employ military forces.

Commanders at the operational level build campaign/major OPLANs to achieve the military strategic/operational objectives. A campaign plan is defined in JP 1-02 as a joint OPLAN for a series of related major operations aimed at achieving strategic or operational objectives within a given time and space. When building and executing the military campaign/major operation plan, commanders at the operational level must ensure military actions are synchronized with those of other government and nongovernmental agencies and organizations, together with international partners, in order to achieve national strategic objectives.

At the heart of the campaign/major OPLAN are operational objectives and associated conditions and tasks for each tactical action contained in the plan. Until the plan is provided to tactical-level commanders, the operational-level commander monitors the operational environment, makes assessments on observations, and replans and updates the campaign/major OPLAN as needed. Once the

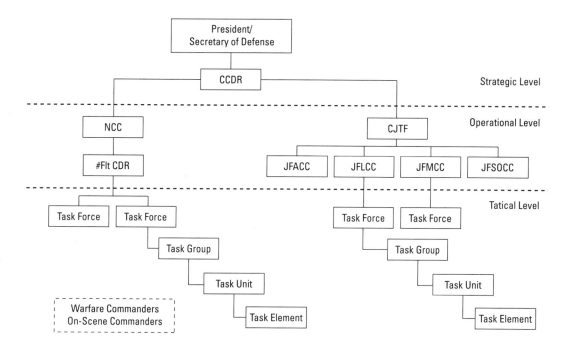

campaign/major OPLAN is provided to tactical-level commanders for execution, the operational-level commander monitors its execution and controls force actions as required.

Tactical Level

Tactics is the employment and ordered arrangement of forces in relation to each other. The tactical level focuses on planning and executing battles, engagements, and activities to achieve military objectives assigned to tactical units or Navy task forces (TFs), task groups (TGs), task units (TUs), and task elements (TEs). An engagement normally is of short duration and includes a wide variety of actions between opposing forces. A battle consists of a set of related engagements, which typically last longer than engagements involving larger forces, such as fleets, armies, and air forces; and normally affect the course of a campaign. Forces at this level generally employ various tactics to achieve their military objectives.

Commanders at the tactical level derive tasking from orders guided by the campaign/major OPLAN. This tasking will include tactical objectives. Tactical objectives often are associated with the specific "target" of an action and are associated with the enemy's centers of gravity (COGs), critical vulnerabilities, and decisive points (DPs). In this context, an objective could be a terrain feature, such as attainment of maritime superiority in a decisive area, the seizing or defending of which is essential to the commander's plan. The objective also could be an enemy force or capability, the destruction of which creates a vulnerability for the adversary.

Maritime Operations Across the Range of Military Operations

Expanding webs of political, military, economic, social, informational, and infrastructure systems provides opportunities for regional powers to compete on a broader scale and emerge on the global landscape with considerable influence. Littoral and urban environments and other complex terrain will increasingly characterize areas of operation that may include humanitarian crisis conditions and combat operations.

Adaptive adversaries continually will seek new capabilities and new employment methods to counter the United States and its allies. As new capabilities, or new methods of employing capabilities, are developed and become more accessible to more players, the conduct of warfare and crisis resolution will change. The nature of war will remain a violent clash of wills between states or armed groups pursuing advantageous political ends. The conduct of warfare will include combinations of conventional and unconventional, kinetic and nonkinetic, and military and nonmilitary actions and operations, all of which add to the increasing complexity of the operational environment.

Future adversaries may lack the ability or choose not to oppose the United States through traditional military action. These adversaries will challenge the United States and its multinational partners by adopting and employing asymmetric methods across selected air, land, maritime, and space domains as well as the information environment against areas of perceived U.S. vulnerability. Many will act and operate without regard for the customary law of war.

Regardless of the type of operation, the Navy and joint forces will require capabilities and processes to respond in the most efficient manner and to minimize the use of military force to that necessary to achieve the overarching strategic objective. This includes the need for engagement before and after conflict/crisis response, the need for integrated involvement with interagency and multinational partners, and the need for multipurpose capabilities that can be applied across the range of military operations.

A common thread throughout the range of military operations is the involvement of a large number of agencies and organizations—many with indispensable practical competencies and significant legal responsibilities and authorities—that interact with the Navy and our multinational partners.

Major Operations or Campaigns Involving Large-Scale Combat

Major operations or campaigns involving large-scale combat place the United States in a wartime state. In such cases, the general goal is to prevail against the enemy as quickly as possible, conclude hostilities, and establish conditions favorable to the United States and its multinational partners.

Crisis Response or Limited Contingency Operation

Crisis response or limited contingency operations can be a single small-scale, limited-duration operation or a significant part of a major operation of extended duration involving combat. The associated general strategic and operational objectives are to protect U.S. interests and prevent surprise attack or further conflict.

The level of complexity, duration, and resources depends on the circumstances. Many of these operations involve a combination of military forces and capabilities in close cooperation with other government agencies (OGAs), IGOs, and NGOs. A crisis may prompt the conduct of foreign humanitarian assistance (FHA), civil support (CS), noncombatant evacuation operations (NEOs), peace operations (PO), strikes, raids, or recovery operations.

MILITARY OPERATIONS OTHER THAN WAR

In contrast to large-scale sustained combat operations, MOOTW focuses on deterring war, resolving conflict, promoting peace, and supporting civil authorities in response to domestic crises. The Marine Corps has a long history of successful participation in MOOTW, from restoring order and nation building in Haiti and Nicaragua from 1900 to the 1930s, to guarding the United States mail in the 1920s. Capturing lessons learned from years of experience in such operations, the Marine Corps published a *Small Wars Manual* in 1940. This seminal reference publication continues to be relevant to today as we face complex and sensitive situations in a variety of operations.

The national security strategy calls for engagement with other nations and a rapid response to political crises and natural disasters to help shape the security environment throughout the world. While this engagement or response may take the form of financial or political assistance, the use of U.S. military forces is always an option for the National Command Authorities. Combatant commanders often rely on responsive, forward-deployed MAGTFs, such as the MEU(SOC), to promote and protect national interests within their area of responsibility. These capable forces, task-organized to meet a variety of contingencies, are usually the first forces to reach the scene and are often the precursor to larger Marine and joint forces.

MOOTW may involve elements of both combat and noncombat operations in peacetime, conflict, and war. Those smaller-scale contingencies involving combat, such as peace enforcement in Haiti in 1995, Operation Urgent Fury in Grenada (1983), Operation El Dorado Canyon in Libya (1986), and Operation Just Cause in Panama (1989), may have many of the same characteristics as war, including offensive and defensive combat operations and employment of the full combat power of the MAGTF. Noncombat operations do not involve the use or threat of force and can help keep the tensions between nations below the threshold of armed conflict or war. In MOOTW, political and cultural considerations permeate planning and execution of operations at all levels of command. As in war, the goal of MOOTW is to achieve national objectives as quickly as possible.

Although normally conducted outside of the United States, MOOTW may be conducted within the United States in support of civil authorities, as demonstrated when Marines assisted civil authorities in restoring order in Los Angeles following the 1992 riots and again following Hurricane Katrina.

The 16 most common missions associated with MOOTW are enumerated below.

Arms Control

Arms control includes those activities by military forces to verify conventional, nuclear, biological, or chemical arms control agreements; seizing or destroying weapons; dismantling or disposing of weapons and hazardous materials; and escorting deliveries of weapons. It encompasses any plan, arrangement or process controlling the numbers, types, and performance characteristics of any weapon system.

Combating Terrorism

Combating terrorism involves actions to oppose terrorism including antiterrorism (defensive measures taken to reduce vulnerability of individuals and property to terrorist acts) and counterterrorism (offensive measures taken to prevent, deter, and respond to terrorism). Marine

Department of Defense Support To Counter Drug Operations

The Department of Defense supports federal, state, and local law enforcement agencies in their efforts to disrupt the transfer of illegal drugs into the United States. Military forces assist in detecting and monitoring drug trafficking; support interdiction efforts; provide intelligence and logistic support; and integrate command, control, communications, computer, and intelligence assets dedicated to interdicting the movement of illegal drugs into the United States.

Enforcement of Sanctions/Maritime Intercept Operations

These operations employ coercive measures to interdict the movement of designated items in or out of a nation or specified area, to compel a country or group to conform to the objectives of the nation or international body that establishes the sanctions. These operations allow authorized cargo and persons to pass through while preventing sanctioned material and persons from entering, as demonstrated during the naval blockade against Iraq throughout the 1990s.

Enforcing Exclusion Zones

These operations employ coercive measures to prohibit specified activities in a specific geographic area. An exclusion zone is established by a sanctioning body to persuade a nation or a group to modify their behavior to meet the desires of the sanctioning body or face continuing sanctions or the use or threat of force. Marine Corps forces, including Marine aviation, can enforce exclusion zones, as demonstrated in Operation Southern Watch (Iraq) and Operation Deny Flight (Bosnia).

Ensuring Freedom of Navigation and Overflight

These operations are conducted to demonstrate U.S. or international rights to navigate sea or air routes in accordance with international law. In 1986 U.S. forces conducted Operation Attain Document, a series of freedom of air and sea navigation operations, against Libya in the Gulf of Sidra. Naval forces can conduct operations to seize and control critical chokepoints on sea lines of communication to help ensure unimpeded use of the seas. Naval aviation can assist the combatant commander provide combat air patrol or strikes against hostile antiair missiles and guns to ensure use of air routes in accordance with international law.

Humanitarian Assistance

Humanitarian assistance operations relieve or reduce the results of natural or manmade disasters that might present a serious threat to life or result in extensive damage to or loss of property. Humanitarian assistance provided by U.S. forces is generally limited in scope and duration. The assistance provided is designed to supplement or complement the host-nation civil authorities efforts. The U.S. military provides assistance when the relief need is gravely urgent and when the humanitarian emergency over-whelms the ability of normal relief agencies to effectively respond. Humanitarian assistance operations may be directed by the National Command Authorities when a serious international situation threatens the political or military stability of a region considered of interest to the U.S. or when the humanitarian situation itself may be sufficient and appropriate for employment of U.S. forces.

The Department of State or the United States ambassador in the affected country is responsible for declaring a foreign disaster or situation that requires humanitarian assistance. The Department of State then requests Department of Defense assistance from the National Command Authorities. Humanitarian assistance operations may cover a broad range of missions. A humanitarian assistance mission could also include securing an environment to allow humanitarian relief efforts.

In 1991, 24th MEU(SOC) provided security, shelter, food, and water to the dissident Kurdish minority in northern Iraq. U.S. military forces participate in three basic types of humanitarian assistance operations: those coordinated by the United Nations, those where the United States acts in concert with other multinational forces or those where the United States responds unilaterally. Naval forces can respond rapidly to emergencies or disasters and achieve order in austere locations. This response could include providing security, logistics, engineering, medical support, and command and control and communications capabilities.

Marine Corps forces can provide sea-based humanitarian assistance. The 5th MEB, during Operation Sea Angel in 1991, assisted Bangladesh in the aftermath of a devastating tropical cyclone by distributing food and medical supplies and repairing the country's transportation infrastructure.

Military Support To Civil Authorities

These operations provide temporary military support to domestic civil authorities when permitted by law and are normally undertaken when an emergency overwhelms the capabilities of the civil authori-ties. Support to civil authorities may be as diverse as temporary augmentation of government workers during strikes, restoration of law and order in the aftermath of riots or protecting life and property, and providing humanitarian relief after a natural disaster.

Limitations on military forces providing support to civil authorities include the Posse Comitatus Act, which prohibits the use of federal military forces to enforce or otherwise execute laws unless expressly authorized by the Constitution or by an act of Congress.

Nation Assistance/Support To Counterinsurgency

Nation assistance is civil or military assistance, other than humanitarian assistance, rendered to a nation by military forces to promote sustained development and growth of civil institutions to promote long-term regional stability. Nation assistance normally consists of security assistance, foreign internal defense, and humanitarian and civic assistance, and is integrated into the U.S. ambassador's country plan. Security assistance is a series of programs the United States provides to defense equipment, military training, and other defense-related services to foreign nations in furtherance of U.S. national policies and objectives. Department of the Navy supports these programs by deploying mobile training teams to conduct military-to-military training. The mission of these teams is to train host-nation personnel to operate, maintain, and employ weapons and support systems or to develop a self-training capability in a particular skill. Teams may be tasked to train either military or civilian personnel, depending on host-nation requests. Foreign internal defense programs encompass all political, economic, informational, and military support provided by the United States to another nation to assist its fight against subversion and insurgency. These programs may address other threats to a host nation's internal stability, such as civil disorder, illegal drug trafficking, and terrorism.

Noncombatant Evacuation Operations

Noncombatant evacuation operations are the evacuation of noncombatants located in a foreign country who are faced with the threat of hostile or potentially hostile actions. They are normally conducted to evacuate U.S. citizens whose lives are in danger, but may also include the evacuation of U.S. military personnel, citizens of the host country, and third country nationals friendly to the United States, as determined by the Department of State.

The Department of State is responsible for protecting and evacuating American citizens abroad and for guarding their property. The Department of Defense advises and assists the Department of State in preparing and implementing plans for evacuating U.S. citizens. The U.S. ambassador or chief of the diplomatic mission is responsible for preparing emergency action plans that address the military evacuation of U.S. citizens and designated foreign nationals from a country. The conduct of military operations to assist implementation of emergency action plans is the responsibility of the geographic combatant commander.

Peace Operations

Peace operations are conducted in support of diplomatic efforts to establish and maintain peace. These operations include peace enforcement, peacekeeping, and operations in support of diplomatic efforts. Peace operations are conducted under the provisions of the United Nations Charter. The specific United Nations resolution under which a peace operation is conducted may dictate rules of engagement, use of combat power, and type of units deployed. Peace enforcement is the application of military force or the threat of its use, usually based on international authorization or consent, to compel compliance with a generally accepted resolution or sanction.

Unlike peacekeeping, peace enforcement does not require the consent of the states involved or of other parties to the conflict and the intervening force is not necessarily considered impartial.

Such operations are conducted under the mandate of Chapter VII of the United Nations Charter. The purpose of peace enforcement is to maintain or restore peace and support diplomatic efforts to reach a long-term settlement. Peace enforcement operation missions include intervention operations, as well as operations to restore order, enforce sanctions, forcibly separate belligerents, and establish and supervise exclusion zones to establish an environment for truce or cease-fire.

Operations in support of diplomatic efforts are those military actions that contribute to the furtherance of U.S. interests abroad. They include:

Preventive Diplomacy. Preventive diplomacy involves diplomatic actions taken in advance of a predictable crisis to prevent or limit violence. Military activities that support preventive diplomacy include deployment of military forces, presence forces, and increased readiness levels to show U.S. resolve and ability to use force to preserve the peace.

Peacemaking. Peacemaking is the process of diplomacy, mediation, negotiation, or other forms of peaceful settlement that arranges an end to a dispute, and resolves issues that led to conflict. Military activities that support peacemaking include security assistance and military-to-military relations.

Peace Building. Peace building consists of post-conflict activities, primarily diplomatic, that strengthen and rebuild civil infrastructure and institutions in order to prevent a return to conflict. These activities include restoring civil authority, rebuilding physical infrastructure, and reestablishing civil institutions like schools and medical facilities.

Protection of Shipping

When necessary, Naval forces can protect U.S. flag vessels, citizens, and their property embarked on United States or foreign vessels from unlawful violence in and over international waters. Protection can take the form of embarked Marines to provide security on board the vessel, ship and convoy escort and CAP by aviation, and the recovery of hijacked vessels by Marine Corps security forces. MAGTFs can also protect U.S. shipping by eliminating threats such as a land-based antiship missiles, coastal guns, and hostile Naval forces operating in the littorals.

Recovery Operations

Recovery operations are conducted to search for, locate, identify, rescue, and return personnel or human remains, sensitive equipment or items critical to national security. Hostile forces may oppose recovery operations. An example of a tactical recovery of aircraft and personnel is the rescue by an MEU of a downed U.S. Air Force pilot in Bosnia in 1995.

REVIEW QUESTIONS

1. Describe the strategic level of warfare.
2. Describe the operational level of warfare.
3. Describe and provide examples of the tactical level of warfare.
4. Describe the primary categories within the Range of Military Operations and provide examples of each.

CHAPTER 3

Joint Organization and Maritime Operations

LEARNING OBJECTIVES

At the end of this chapter the student will be able to:

- Explain the role of the National Command Authorities.
- Explain the functions of the Department of Defense.
- Explain the common functions of the military departments.
- Explain the functions of the Department of the Navy and the Marine Corps.
- Explain the functions of the Chairman of the Joint Chiefs of Staff.
- Describe why a Joint Staff is essential.
- Explain the difference between a unified and specified combatant command.
- Explain the difference between combatant commanders assigned geographic area responsibilities and combatant commanders assigned functional responsibilities.

Introduction to Joint Organization

Joint Organization before 1900. As established by the Constitution, coordination between the War Department and Navy Department was effected by the President as the Commander in Chief. **Army and naval forces functioned autonomously** with the President as their only common superior. However, instances of confusion, poor inter-Service cooperation and lack of coordinated, joint military action had a negative impact on operations in the Cuban campaign of the Spanish-American War (1898). By the turn of the century, advances in technology and the growing international involvement of the United States required greater cooperation between the military departments.

Joint history through World War I. As a result of the unimpressive joint military operations in the Spanish-American War, in 1903, the Secretary of War and the Secretary of the Navy created the Joint Army and Navy Board charged to address **"all matters calling for cooperation of the two Services."** The Joint Army and Navy Board was to be a continuing body that could plan for joint operations and resolve problems of common concern to the two Services. Unfortunately, the Joint Board accomplished little, because it could not direct implementation of concepts or enforce decisions, being limited to

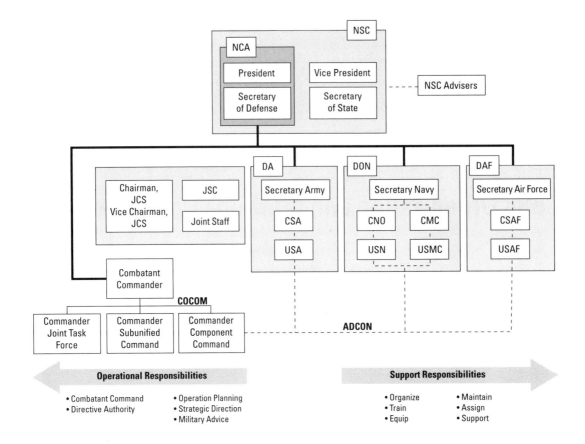

commenting on problems submitted to it by the secretaries of the two military departments. It was described as "a planning and deliberative body rather than a center of executive authority." As a result, it had little or no impact on the conduct of joint operations during the first World War. Even as late as World War I, questions of seniority and command relationships between the Chief of Staff of the Army and American Expeditionary Forces in Europe were just being resolved.

Joint History through World War II. Soon after the Pearl Harbor attack, President Roosevelt and Prime Minister Churchill met with their military advisers at the Arcadia Conference in Washington to plan a coordinated effort against the Axis powers. At that time, the two allied leaders established the Combined Chiefs of Staff (CCS) as the supreme military body for strategic direction of the Anglo-American war effort. British representation for the new organization consisted of the Chiefs of Staff Committee composed of the heads of the British armed services who had been giving effective administrative coordination, tactical coordination, and strategic direction to British forces for almost 20 years. The British committee served as a "corporate" body for giving military advice to the War Cabinet and the Prime Minister. The collective responsibility of the British committee was set by the Prime Minister in 1924 and given to each new member as a directive:

> In addition to the functions of the Chiefs of Staff as advisers on questions of sea, land or air . . . each of the three Chiefs of Staff will have an individual and collective responsibility for advising on defense policy as a whole, the three constituting, as it were, a Super-Chief of a War Staff in Commission.

But the United States in 1941 had no established agency to furnish U.S. input to a Combined Chiefs of Staff committee. Consequently, the U.S. officers whose positions and duties matched those of

FUNCTIONS OF THE DEPARTMENT OF DEFENSE

As prescribed by the National Security Act of 1947, as amended, the Department of Defense maintains and employs the Armed Forces to

- support and defend the Constitution of the United States against all enemies, foreign and domestic;
- ensure, by timely and effective military action, the security of the United States, its possessions, and areas vital to its interest; and
- uphold and advance the national policies and interests of the United States.

the British Chiefs of Staff committee formed the U.S. position of the CCS; that group became known as the Joint U.S. Chiefs of Staff. This first Joint Chiefs of Staff worked throughout the war without legislative sanction or even formal Presidential definition, a role that President Roosevelt believed preserved the flexibility required to meet the needs of the war. The initial members of the Joint U.S. Chiefs of Staff were Admiral William D. Leahy, President Roosevelt's special military adviser, with a title of Chief of Staff to the Commander in Chief of the Army and Navy; General George C. Marshall, Chief of Staff of the Army; Admiral Ernest J. King, Chief of Naval Operations and Commander in Chief of the U.S. Fleet; and General Henry H. Arnold, Deputy Army Chief of Staff for Air and Chief of the Army Air Corps.

Under President Roosevelt's leadership, this new U.S. military body steadily grew in influence and became the primary agent in coordinating and giving strategic direction to the Army and Navy. In combination with the British Chiefs of Staff, it mapped and executed a broad strategic direction for both nations.

At the end of World War II, the continued need for a formal structure of joint command was apparent; the wartime Joint Chiefs of Staff offered an effective workable example. The first legislative step was the passage of the National Security Act of 1947, which formally established the Joint Chiefs of Staff and laid the foundation for the series of legislative and executive changes that produced today's defense organization.

Since passage of the National Security Act of 1947, the President has used his Secretary of Defense as his **principal assistant** in all matters relating to the Department of Defense. The Secretary is responsible for the effective, efficient, and economical operation of the Department of Defense, and he has statutory authority, direction, and control over the military departments.

Organization For National Security.

Knowledge of relationships between elements of the national security structure is essential to understanding the role of joint staff organizations.

NATIONAL COMMAND AUTHORITIES (NCA)

Constitutionally, the ultimate authority and responsibility for the national defense rests with the President.

The **National Command Authorities (NCA)** are the President **and** Secretary of Defense or persons acting lawfully in their stead. The term NCA is used to signify constitutional authority to direct the armed forces in their execution of military action. Both movement of troops and execution of military action must be directed by the NCA; by law, no one else in the chain of command has the authority to take such action except in self-defense.

National Security Council (NSC). The National Security Council was established by the National Security Act of 1947 as the principal forum to consider national security issues that require Presidential decision. Its membership now includes only four **statutory members**: the President, the Vice President, the Secretary of State, and the Secretary of Defense. The Chairman of the Joint Chiefs of Staff (CJCS) and the Director of Central Intelligence serve as **statutory advisers** to the NSC.

The role of the **Secretary of Defense** has significantly changed since the position was established in 1947. Originally, the secretary had only general authority shared with the civilian secretaries of the military departments. Subsequent legislation incrementally strengthened the Secretary of Defense's authority. Today the Secretary of Defense is the principal assistant to the President for all matters relating to the Department of Defense.

This diagram illustrates the organization that reports to the Secretary of Defense.

DOD ORGANIZATION (JUNE 2000)

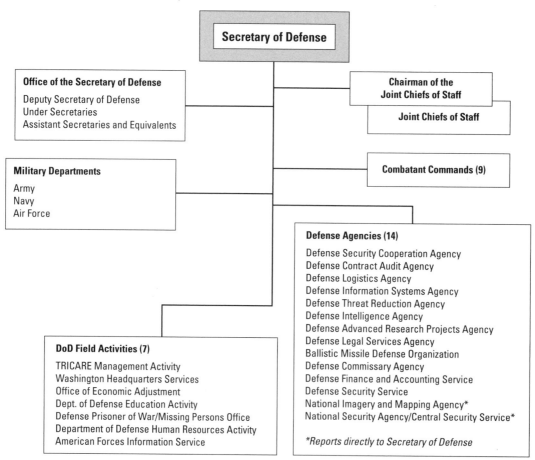

Geographic Combatant Commands	Functional Combatant Commands
U.S. Africa Command	U.S. Joint Forces Command
U.S. Central Command	U.S. Special Operations Command
U.S. European Command	U.S. Strategic Command
U.S. Northern Command	U.S. Transportation Command
U.S. Pacific Command	
U.S. Southern Command	

Geographic and Functional Combatant Commands as of June 2008

Chairman of the Joint Chiefs of Staff (CJCS). The Goldwater-Nichols DOD Reorganization Act of 1986 identified the CJCS as the head of the Joint Chiefs of Staff and the senior ranking member of the Armed Forces. By law, CJCS is now the **principal** military adviser to the President. As appropriate, the CJCS may seek the advice of and consult with the other JCS members and combatant commanders. When CJCS presents advice, he presents the advice or opinions of other JCS members and, as he considers appropriate, the range of military advice and opinions he has received.

    Vice Chairman of the Joint Chiefs of Staff (VCJCS). The DOD Reorganization Act of 1986 created the position of Vice Chairman of the Joint Chiefs of Staff, who performs such duties as the CJCS may prescribe. By law, VCJCS is the second ranking member of the armed forces. In the absence or disability of CJCS, the Vice Chairman acts as, and performs the duties of, the Chairman. Though not originally included as a member of the JCS, VJCS was vested by Section 911 of the National Defense Authorization Act of 1993 as a full voting member of the Joint Chiefs of Staff.

    Military Service Chiefs. The military Service chiefs "wear two hats." As the chiefs of the military Services, they perform their duties under the authority, direction, and control of the secretaries of the military departments and are directly responsible to their Service secretaries. As members of the Joint Chiefs of Staff, they offer **advice** to the President, Secretary of Defense, and NSC. By custom, the vice chiefs of the Services are delegated authority to act for their chiefs in most matters having to do with day-to-day operation of the Services.

Combatant Commands

National Security Act (NSA) of 1947 was the first definitive legislative statement "to provide for the effective strategic direction of the armed forces and for their operation under unified control and for their integration into an efficient team of land, naval, and air forces." The act went on to say that it was the responsibility of the Joint Chiefs of Staff to "establish unified commands in strategic areas when such unified commands are in the interest of national security," and the President would establish unified and specified combatant commands to perform military missions. The military departments would assign forces to the combatant commands; the responsibility for their support and administration would be assigned by the Secretary of Defense to a military department. Forces not assigned would remain under the authority of the military department.

Unified Combatant Command. A military command which has a **broad, continuing mission** under a single commander and which is composed of forces from **two or more military departments**.

Specified Combatant Command. A military command which has a broad, continuing mission and which is **normally** composed of forces from **one military department**. There are currently no specified commands but the option to create such a command still exists.

The commanders of combatant commands exercise combatant command (command authority) **(COCOM)** of assigned forces and are directly responsible to the NCA for the performance of assigned missions and the preparedness of their commands.

Military Departments

The chain of command for purposes other than the operational direction of combatant commands runs from the President to the Secretary of Defense to the secretaries of the military departments to the chiefs of the Service forces. The **military departments** are separately organized, each under civilian secretaries who are responsible for, and have the authority to conduct all affairs of their respective departments, including the following:

- recruiting
- supplying
- training
- mobilizing
- administering
- construction, outfitting, and repairing equipment
- acquisition of real property

- organizing
- equipping
- servicing
- demobilizing
- maintaining
- construction, maintenance, and repair of buildings, structures, and utilities

The U.S. Army

ORIGIN

From its birth in 1775 until the early 1800s, young America's army staff patterned itself after the British system: control of the small Regular Army was split between the Commanding General, who was responsible for military discipline and control of field forces, and the Secretary of War, who guided administration and support with a staff bureau system. This bureau system divided authority between the Secretary of War and the Commanding General of the Army and lacked the mechanism to develop coordinated, long-range plans. Though suited to the efficient administration of a small peacetime force, the bureau system was incapable of coping with the demands placed on the twentieth-century Army, a situation that became clear in the Spanish-American War (1898).

DEVELOPMENT IN THE TWENTIETH CENTURY

In 1899, a civilian lawyer, Elihu Root, was appointed Secretary of War. At the time, he expanded the Army's missions to include pacification and administration of the island territories recently acquired from Spain; in addition, he responded to public criticism of the logistical and operational confusion that had plagued Army performance in the Spanish-American War. He undertook reform of the Army command and staff system patterned on the British system. In 1903 Congress passed legislation creating a modern U.S. Army General Staff. The War Department General Staff corps of 44 officers, who were relieved of all other duties, was functionally organized to prepare plans for the national defense and mobilization of troops. The legislation also replaced the ranking military position, Commanding General of the Army, with a War Department Chief of Staff. The Chief of Staff (COS) supervised all Army forces and the staff departments that had been responsible to the Secretary of War. It was not until 1918,

FUNCTIONS OF THE DEPARTMENT OF THE ARMY

The Army is responsible for the preparation of land forces necessary for the effective prosecution of war and military operations short of war, and, in accordance with integrated joint mobilization plans, for the expansion of the peacetime components of the Army to meet the needs of war. The Army, within the Department of the Army, includes land combat and service forces and any organic aviation and water transport assigned.

Some of the Major Functions of the Army Are to:
- organize, train, and equip forces for the conduct of prompt and sustained combat operations on land—specifically, forces to defeat enemy land forces and to seize, occupy, and defend land areas;
- organize, train, equip, and provide forces for appropriate air and missile defense and space control operations, and for the support and conduct of special operations;
- develop airborne doctrine, procedures, and equipment that are of common interest to Army and Marine Corps;
- organize, equip, and provide Army forces for joint amphibious, airborne, and space operations and train such forces, in accordance with joint doctrines;
- organize, equip, and provide forces for the support and conduct of special operations;
- organize, equip, and provide forces for the support and conduct of psychological operations;
- furnish forces for the occupation of territories abroad; and
- conduct the authorized civil works program, including projects for improvement of navigation, flood control, beach erosion control, and other water resource developments in the United States.

A collateral function of the Army is to train forces to interdict enemy sea and air power and communications through operations on or from land.

though, that it was clearly resolved that the Chief of Staff was the ranking member of the Army when General Pershing, then Commander of the American Expeditionary Force, was made subordinate to the COS. The Root reforms were the beginning that gave the Army the basis for a unified command and staff system.

Today the **Army Staff** is an executive component of the Department of the Army. It exists to assist the Secretary of the Army in his/her responsibilities, and includes the following:

- Chief of Staff
- Vice Chief of Staff
- Deputy Chiefs of Staff for Personnel, Intelligence, Operations and Plans, and Logistics
- Assistant Chiefs of Staff (positions authorized by law, but not used) Special Staff: Chief of Engineers; Surgeon General; Judge Advocate General; Chief of Chaplains; Chief of National Guard Bureau; and Chief of Army Reserves

FUNCTIONS OF THE DEPARTMENT OF THE NAVY

The Department of the Navy is responsible for the preparation of the Navy and Marine Corps forces necessary for the effective prosecution of war and military operations short of war and, under the integrated joint mobilization plans, for the expansion of the peacetime component of the Navy and Marine Corps to meet the needs of war. Within the Department of the Navy, the Navy includes naval combat and service forces and such aviation as may be organic.

Some of the Major Functions of the Navy and Marine Corps Are to:

- organize, train, equip and furnish Navy and Marine Corps forces for the conduct of prompt and sustained combat incident to operations at sea, including operations of sea-based aircraft and land-based naval air components—specifically, forces to seek out and destroy enemy naval forces and to suppress enemy sea commerce, to gain and maintain general naval supremacy, to establish and maintain local superiority in an area of naval operations, to seize and defend advanced naval bases, and to conduct such land, air, and space operations as may be essential to the prosecution of a naval campaign;
- organize, equip, and furnish naval forces, including naval close air support and space forces, for the conduct of joint amphibious operations;
- organize, train, equip, and provide forces for strategic nuclear warfare to support strategic deterrence;
- organize, train, equip, and provide forces for reconnaissance, antisubmarine warfare, protection of shipping, aerial refueling and minelaying, and controlled minefield operations; furnish the afloat forces for strategic sealift;
- furnish air support essential for naval operations;
- organize, train, equip, and provide forces for appropriate air and missile defense and space control operations, including forces required for the strategic defense of the United States, under joint doctrines;
- organize, train, equip, and furnish forces to operate sea lines of communication;
- organize, train, equip, and furnish forces for the support and conduct of special operations; and
- coordinate with the Department of Transportation for the peacetime maintenance of the Coast

Some Collateral Functions of the Navy and Marine Corps Are to:

- interdict enemy land power, air power, and communications through operations at sea;
- furnish close air and naval support for land operations;
- prepare to participate in the overall air and space effort; and
- establish military government pending transfer of this responsibility.

The U.S. Navy

ORIGIN

The Department of the Navy was established in 1798. The early department was entirely in the hands of civilian appointees, while naval officers served at sea. Growth in size and complexity of Navy business in the first quarter of the 1800s led to creation of a Board of Naval Commissioners to give professional advice to the civilian appointees on constructing, repairing, and equipping ships and superintending shipyards. It was a bilinear arrangement, since employment of forces and discipline of troops were retained by the Secretary of the Navy. By 1842 the Navy Department had shifted from a predominantly personnel service, like its Army counterpart, to a predominantly materiel service deeply involved in complex and expanding technical problems. Five individual bureaus under the Secretary of the Navy were created for yards and docks; construction, equipment, and repairs; provisions and clothing; ordnance and hydrography; and medicine and surgery. The creation of additional bureaus specifically for navigation and equipment and for recruiting (enlisted personnel matters) was the response to weaknesses of the bureau system that were discovered during the Civil War. When necessary, special boards were formed to consider specific technical problems, such as strategy, inventions, and new vessels. By the close of the nineteenth century, the size and complexity of the Service, as well as the pressing need to ensure adequate preparation for war, became too much for control by a single manager. This, compounded by the intra-Service as well as the inter-Service experiences in the Spanish-American War, furnished motivation for Congressional and administrative change in the early 1900s.

DEVELOPMENT IN THE TWENTIETH CENTURY

In 1909 a General Board of the Navy was established to serve as an advisory body to the secretary on matters of personnel, operations, materiel, and inspections. Legislation in 1915 created the Office of the Chief of Naval Operations (CNO) that was charged with the operation of the fleet and preparation and readiness of war plans. In the 1920s the responsibilities for operation of the fleet were assigned to the newly created position of Commander in Chief of the U.S. Fleet. In March 1942 the positions of Commander in Chief of the U.S. Fleet and CNO were consolidated; once again the total direction and support of the U.S. Navy operating forces were under a single person. By the 1960s the CNO as military chief had complete responsibility for operations as well as supporting logistics and administration.

Today the **Office of the Chief of Naval Operations** within the Department of the Navy assists the Secretary of the Navy in executing his or her responsibilities. This office includes the following:

- Chief of Naval Operations
- Vice Chief of Naval Operations
- Assistant Vice Chief of Naval Operations
- Deputy Chiefs of Naval Operations for Manpower and Personnel (N1); Policy, Strategy, and Plans (N3/5)   Logistics (N4) Resources, Warfare Requirements and Assessments (N8)
- Directors:
    Director of Naval Intelligence (N2)
    Director, Space and Command, Control, Communications, Computers, and Intelligence (C4I) Requirements (N6)
    Director, Training and Doctrine (N7)
    Chief of Naval Reserve; Surgeon General; Chief of Chaplains; and Oceanographer of the Navy.

FUNCTIONS OF THE MARINE CORPS

Specific responsibilities of the Department of the Navy toward the Marine Corps include the maintenance of not less than three combat divisions and three air wings and such other land combat, aviation, and other services as may be organic therein.

Some of the Major Functions of the Marine Corps Are to:

- organize, train, and equip Fleet Marine Forces of combined arms, together with supporting air components, for service with the fleet in the seizure or defense of advanced naval bases and for the conduct of such land operations as may be essential to the prosecution of a naval campaign;
- furnish security detachments and organizations for service on naval vessels of the Navy;
- furnish security detachments for protection of naval property at naval stations and bases;
- perform other duties as the President may direct; and
- develop landing force doctrines, tactics, techniques, and equipment that are of common interest to the Army and Marine Corps.

The U.S. Marine Corps

ORIGIN

The Marine Corps staff had its origin in 1798 in the Act for the Establishment and Organization of the Marine Corps. For a time the Commandant was a one-man staff; his chief duty was recruiting Marines for service with the fleet. As the number of recruits began to increase, however, the Commandant expanded the staff to include an adjutant to assist with musters and training, a quartermaster to procure supplies, and a paymaster to pay the troops. An administrative staff of three to five officers carried the Marine Corps through the nineteenth century.

STAFF GROWTH IN THE TWENTIETH CENTURY

The emergence of the United States as a world power after the Spanish-American War greatly expanded Marine Corps employment. As additional staff officers were assigned to aid the adjutant, quartermaster, and paymaster, their offices became known as departments. Change first occurred outside the staff departments in what came to be called the "Immediate Office of the Commandant." The initial step was taken in 1902, when an officer was assigned to headquarters as aide-de-camp to the Commandant. He formed the nucleus for staff expansion in the Office of the Commandant. The position of Chief of Staff was added in 1911 to assist the Commandant with matters of training, education, equipping the troops, and organization, distribution, and assembly at embarkation for expeditionary duty.

Between World War I and the 1970s, the Marine Corps headquarters staff evolved into the staff that is seen today. In the early years of the twentieth century, there was the strong influence of the American Expeditionary Force and the development of the Army staff. Through World War II, the headquarters

FUNCTIONS OF THE DEPARTMENT OF THE AIR FORCE

The Department of the Air Force is responsible for the preparation of the air forces necessary for the effective prosecution of war and military operations short of war and, under integrated joint mobilization plans, for the expansion of the peacetime component of the Air Force to meet the needs of war. Within the Department of the Air Force, the Air Force includes combat and service aviation forces.

Some of the Major Functions of the Air Force Are to:
- organize, train, equip, and provide forces for the conduct of prompt and sustained combat operations in the air–specifically, forces to defend the United States against air attack, gain and maintain general air supremacy, defeat enemy air forces, conduct space operations, control vital air areas, and establish local air superiority;
- organize, train, equip, and provide forces for appropriate air and missile defense and space control operations, including forces for the strategic defense of the United States, in accordance with joint doctrines;
- organize, train, equip, and provide forces for strategic air and missile warfare;
- organize, equip, and provide forces for joint amphibious, space, and airborne operations;
- organize, train, equip, and provide forces for dose air support and air logistic support to the Army and other forces, including airlift, air support, resupply of airborne operations, aerial photography, tactical air reconnaissance, and air interdiction of enemy land forces and communications;
- organize, train, equip, and provide forces for air transport for the armed forces;
- develop doctrines, procedures, and equipment for air defense from land areas;
- furnish launch and space support for the Department of Defense;
- organize, train, equip, and furnish land-based tanker forces for the in-flight refueling support of strategic operations and deployments of aircraft of the Armed Forces and Air Force tactical operations;
- organize, train, equip, and furnish forces to operate air lines of communications; and
- organize, train, equip, and furnish forces for the support and conduct of special operations.

Collateral Functions of the Air Force include:
- surface sea surveillance and antisurface ship warfare through air operations,
- antisubmarine warfare and antiair warfare operations to protect sea lines of communications,
- aerial minelaying operations; and
- air-to-air refueling in support of naval campaigns.

staff retained a line planning staff and functionally organized staff divisions for administrative, technical, supply, and operations functions. In the 1950s the staff was reorganized along general staff divisions, G-1 through G-4, and several technical staff divisions. The position of Chief of Staff was redefined in 1957 to assist the Commandant in his responsibilities to supervise and coordinate the headquarters staff. Even through the early 1970s, there was a composite staff arrangement with a distinction in line and staff functions. In 1973 headquarters was reorganized along functional lines with four Deputy Chiefs of Staff: Manpower, Installations and Logistics, Requirements and Programs, and Plans and Operations. These new directorates replaced the general staff sections. Marine Corps field units continued to use a combination of a functionally organized general and executive staff and a staff of technical experts.

The **Headquarters, Marine Corps**, is in the executive part of the Department of the Navy. Its functions are to furnish professional assistance to the Secretary of the Navy, accomplish all military department support duties that deal with the Marine Corps, coordinate the action of Marine Corps organizations, prepare instructions for the execution of approved plans, and investigate and report efficiency of the Marine Corps in support of combatant commands. Its current organization includes the following:

- Commandant of the Marine Corps
- Assistant Commandant of the Marine Corps
- Director Marine Corps Staff
- Deputy Commandant for:
    Aviation
    Installation and Logistics
    Manpower and Reserve Affairs
    Plans, Policies and Operations
    Programs and Resources
- Assistant Commandant for Command, Control, Communications, Computers, and Intelligence (C4I)

The U.S. Air Force

ORIGIN

The earliest staff organization in the Air Force reflected the general staff organization in the Army in the years before World War II. Before 1935 the War Department General Staff was responsible for planning, coordinating, and controlling the Air Corps. In 1935 the General Headquarters Air Force was formed and operated under the Army Chief of Staff and the War Department. By June 1941 the Army Air Forces had a recognized Office of the Chief of the Air Force. Reorganization throughout the war years resulted in experiments with a variety of staff organizational arrangements: the Army-style general staff organization; a double-deputy staff that produced a two-prong functional general staff identified as operations and administration; and a tridirectorate staff that recognized personnel and administration, materiel and logistics, and plans and operations.

GROWTH SINCE 1947

With the passage of the National Security Act of 1947, the U.S. Air Force was created as a separate military Service and a coequal partner in the National Military Establishment. At first, the U.S. Air Force retained the multiple directorate organization used when it was the Army Air Corps. The first Secretary of the Air Force was sworn in on 18 September 1947. The Secretary, along with the first several Chiefs of Staff, developed what was to become the foundation of today's headquarters staff. The current organization is a multiple directorate staff: the traditional personal and specialist staff subdivisions plus a coordinating staff of personnel, comptroller, operations, and materiel.

FUNCTIONS OF THE COAST GUARD

The Coast Guard is a military Service and a branch of the Armed Forces of the United States at all times. It is a Service in the Department of Transportation except when operating as part of the Navy on declaration of war or when the President directs.

Some of the Major Peacetime Functions of the Coast Guard Are to:

- enforce or assist in enforcement of the law with power to arrest, search, and seize persons and property suspected of violations of Federal law, including drug interdiction;
- administer laws and enforce regulations for the promotion of safety of life and property on and under the high seas and waters subject to U.S. jurisdiction;
- coordinate marine environmental protection response;
- enforce port safety and security;
- enforce commercial vessel safety standards and regulations;
- regulate and control ship movement and anchorage;
- acquire, maintain, and repair short-range aids to navigation;
- establish, operate, and maintain radio navigation;
- develop, establish, maintain, and operate polar and U.S. icebreaking facilities;
- organize, equip, and furnish forces for maritime search and rescue;
- engage in oceanographic research; and
- maintain a state of readiness to function as a specialized Service in the Navy.

Some of the Major Wartime Functions of the Coast Guard Are to:

- continue peacetime missions;
- plan and coordinate U.S. coastal defense for the Fleet Commanders through assignment as commanders of U.S. Maritime Defense Zone Atlantic and Pacific; and
- perform naval wartime missions of inshore undersea warfare, mine countermeasures, harbor defense, ocean escort, etc., occurring in the U.S. littoral sea.

Since its inception, the U.S. Air Force has been organized along functional rather than area lines. The Chief of Staff is the military head of the Air Force. The Deputy Chiefs of Staff may speak for the Chief of Staff at any time on any subject within their functional areas, according to the authority delegated by the Chief of Staff. Each deputy in turn presides over a family of directorates, and each directorate is functionally oriented. In the Air Staff, decisions are made at the lowest level that has access to sufficient information and the requisite delegated authority.

The **Air Staff** is an executive part of the Department of the Air Force. It serves to assist the Secretary of the Air Force in carrying out his responsibilities and is organized as follows:

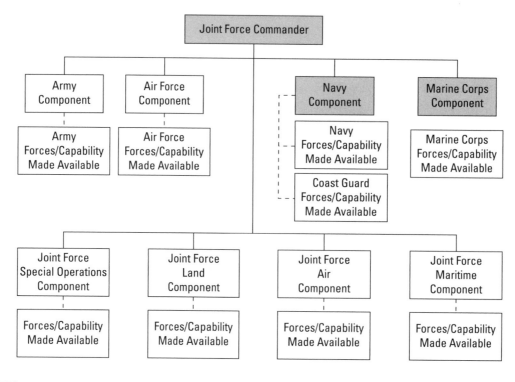

NOTES:

(1) A joint force contains Service components (becasue of logistic and training responsibilities), even when operations are conducted through functional components.

(2) All Service and functional components are depicted; and mix of the above components can constitute a joint force.

(3) There may also be a Coast Guard component in a joint force.

LEGEND:

Operational Control (OPCON) ——————

Command Relationship Determined by JFC - - - - - - - - -

- Chief of Staff of the Air Force
- Vice Chief of Staff
- Deputy Chiefs of Staff for:
    Personnel
    Installations and Logistics
    Plans and Programs
    Air and Space Operations
    Director of Headquarters
    Communications and Information
- Assistant Chief of Staff for Intelligence
- Special Staff:

Surgeon General	Chief of National Guard Bureau
Judge Advocate General	Chief of Safety
Chief of Chaplains	Director of Manpower and Organization
Chief of Security Police	Director of Programs and Evaluation
Director of Test and Evaluation	Civil Engineer
Chief of Air Force Reserve	Air Force Historian
Director of Morale, Welfare, Recreation and Services	

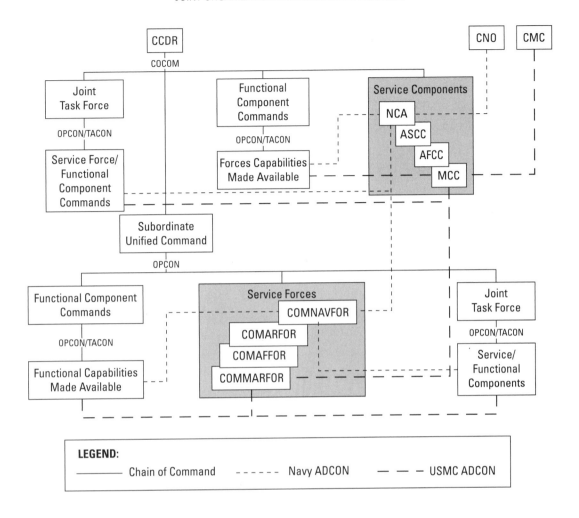

LEGEND:
——— Chain of Command ------ Navy ADCON — — — USMC ADCON

The U.S. Coast Guard

ORIGIN

The Coast Guard, the nation's oldest continuing seagoing Service, was established in 1790 as "a system of cutters" in the Treasury Department. First called the Revenue Marine and later the Revenue Cutter Service, the Coast Guard was primarily a law enforcement agency responsible for collecting customs duties from ships entering U.S. waters, enforcing embargoes, hunting pirates, and enforcing quarantines. However, by 1797 the strength of the Treasury Department's cutters had been increased to "defend the sea coast and repel any hostility to vessels and commerce"; Congressional authorization established the role of the Coast Guard in national defense.

EXPANSION OF RESPONSIBILITY

In 1915 the U.S. Lifesaving Service, an organization of local stations scattered along U.S. coasts, merged with the Revenue Cutter Service to form the U.S. Coast Guard, and with that was born its traditional image, the "lifesavers." During World War I responsibilities were added for port safety and security, commercial vessel safety, icebreaking, and marine environment protection. Joined in 1939 by the Lighthouse Service, the Service assumed responsibility for establishing and maintaining aids to navigation. In 1967 the Coast Guard became part of the newly formed Department of Transportation. A comprehensive review of wartime missions was performed in 1981 by the Navy and Coast Guard Board.

In a 1984 Memorandum of Understanding between the Secretaries of Navy and Transportation, Coast Guard area commanders were assigned as commanders of the newly formed U.S. Maritime Defense Zones (MDZ). These commanders are responsible to the Atlantic and Pacific Fleet commanders for planning and coordinating U.S. coastal defense, preparing operation plans, conducting exercises, and training reserve forces. MDZs will be activated when needed as a deterrent option to ensure port safety and the initial safety of seaborne deployments.

ORGANIZATION

The command and control structure of the Coast Guard is based on nine autonomous districts and two Maintenance and Logistics Commands (MLCs) that report to the Atlantic and Pacific area commanders. The Commandant of the Coast Guard reports directly to the Secretary of Homeland Security in peacetime. On declaration of war, or when directed by the President, the Coast Guard becomes a Service within the Navy with the Commandant reporting to the Secretary of the Navy; he or she reports to the CNO for military functions concerning organization, training, and readiness of operational forces assigned to the Navy.

The **Headquarters, U.S. Coast Guard**, under the Commandant reports in peacetime to the Secretary of Transportation. The Commandant is assisted in the direction of policy, legislation, and administration by a functional organization headed by Chiefs of Offices:

- Chiefs of Offices:

Acquisition	Chief Counsel
Civil Rights	Command, Control, and Communications
Resource Director/Comptroller	Engineering
Health Services	Marine Safety, Security, and Environmental Protection
Navigation	Operations
Personnel	Readiness and Reserves

Joint Maritime Operations

The President, through the Secretary of Defense (SecDef) and with the advice and assistance of the Chairman of the Joint Chiefs of Staff (CJCS), establishes combatant commands for the performance of military missions and prescribes the force structure of such commands. The CCDR, commander subunified command, and JTF commander are joint force commanders (JFCs) who normally exercise OPCON over assigned (and normally over attached) forces. JTFs can be established by the SecDef, a CCDR, or an existing JTF commander.

The CCDR may conduct operations through the Service component commanders; subordinate JFCs may conduct operations through Service force commanders, e.g., Navy TF commanders. This relationship is appropriate when stability, continuity, economy, ease of long-range planning, and the scope of operations dictate organizational integrity of Service forces for conducting operations.

In this figure, levels of command in the joint force that naval commanders routinely occupy are shaded. These roles are JFC, Navy component commander (NCC), Marine component commander (MCC), and joint force maritime component commander (JFMCC).

JOINT FORCE COMMANDER

A commander, joint task force (CJTF) is a joint force commander (JFC). Navy commanders may be designated a JFC by the SecDef, a CCDR, a subordinate unified commander, or an existing CJTF. CJTFs normally are operational-level commanders. The authority that establishes the joint command will state the forces that are made available and also include the overall mission, purpose, and objectives for the directed military operations.

NAVY COMPONENT COMMANDER

A Navy component command assigned to a CCDR consists of the NCC and the Navy forces (NAVFOR) (such as individuals, units, detachments, and organizations, including the support forces) that have been assigned to that CCDR.

A Marine component command assigned to a CCDR consists of the MCC and the Marine forces (such as individuals, units, detachments, and organizations, including the support forces) that have been assigned to that CCDR.

In addition to fulfilling any designated command authorities, Service component commanders have responsibilities that derive from their Services' support function.

NCCs/MCCs are always responsible for the following Navy/Marine Corps–specific functions:

1. Makes recommendations to the JFC on the proper employment of the Navy/Marine Corps forces.
2. Accomplishes such operational missions as may be assigned.
3. Selects and nominates specific units of the Navy/Marine Corps component for attachment to other subordinate commands. Unless otherwise directed, these units revert to the NCC/MCC's control when such subordinate commands are dissolved.
4. Conducts joint training, including, as directed, the training of components of other Services in joint operations for which the NCC/MCC has or may be assigned primary responsibility, or for which the Navy/Marine Corps component's facilities and capabilities are suitable.
5. Informs their JFC (and their CCDR, if affected) of planning for changes in logistic support that would significantly affect operational capability or sustainability. If the CCDR does not approve the proposal and discrepancies cannot be resolved between the CCDR and the NCC/MCC, the CCDR will forward the issue through the CJCS to the SecDef for resolution. Under crisis action or wartime conditions, and where critical situations make diversion of the normal logistic process necessary, NCC/MCCs will implement directives issued by the CCDR.
6. Develop program and budget requests that comply with CCDR guidance on warfighting requirements and priorities.
7. Informs the CCDR (and any intermediate JFCs) of program and budget decisions that may affect joint operation planning.
8. As requested, provides supporting joint operation and exercise plans with necessary force data to support missions that may be assigned by the CCDR.

NCC/MCC or other Navy/Marine Corps commanders assigned to a CCDR are responsible to the CNO/CMC for the following:

1. Internal administration and discipline
2. Training in joint doctrine and Navy/Marine Corps doctrine, tactics, techniques, and procedures
3. Logistic functions normal to the command, except as otherwise directed by higher authority
4. Navy/Marine Corps intelligence matters and oversight of intelligence activities to ensure compliance with the laws, policies, and directives.

Unless otherwise directed by the CCDR, the NCC/MCC will communicate through the combatant command on those matters over which the CCDR exercises COCOM.

In this figure, the Navy administrative control (ADCON) authority always exists from the CNO to the NCC and from the NCC to subordinate COMNAVFORs. Likewise, ADCON authority always exists from the CMC to the MCC and from the MCC to subordinate COMMARFORs.

REVIEW QUESTIONS

1. What is role of the National Command Authorities?
2. List the common functions of the military departments.
3. Explain the roles of the NCC and MCC.
4. Explain the functions of the Chairman of the Joint Chiefs of Staff.
5. Describe why a Joint Staff is essential.
6. What is the difference between a unified and specified combatant command?

CHAPTER 4

Naval Organization and Command and Control

LEARNING OBJECTIVES

At the end of this chapter the student will be able to:

- Explain the origination of the department of the Navy. Understand the administrative and operational Chains of Command for the Navy and Marine Corp.
- Know the difference between Major and Numbered Fleets.
- Understand Task Organization. Define Command and Control and describe the three components which form the basis of the command and control system.
- Explain the purpose for "Rules of Engagement."
- Describe the Composite Warfare Commander (CWC) concept, how and why it works, and the standard organization.
- State the mission and functions of CIC.
- Describe the four conditions of readiness.
- Identify basic NTDS symbology. Interpret the meaning of warning, weapons, and engagement orders.
- Describe the Detect-to-Engage Sequence.

ADDITIONAL READING

Naval Doctrine Publication 6, *Naval Command and Control*. CNO, Washington D.C., 19 May 1996.

NAVEDTRA 10776-A: *Surface Ship Operations,* Naval Education and Training Command Government Printing Office, Washington D.C., SWOSCOLPAC, *Engagement Systems,* Surface Warfare Officers School Division Officer Course, June 1991.

Navy Department Organization

The Department of the Navy has three principal components:

The Navy Department, consisting of executive offices mostly in Washington, D.C.

The operating forces, including the Marine Corps, and the reserve components; and, in time of war, the U.S. Coast Guard (in peace, a component of the Department of Homeland Security).

The shore/support establishment.

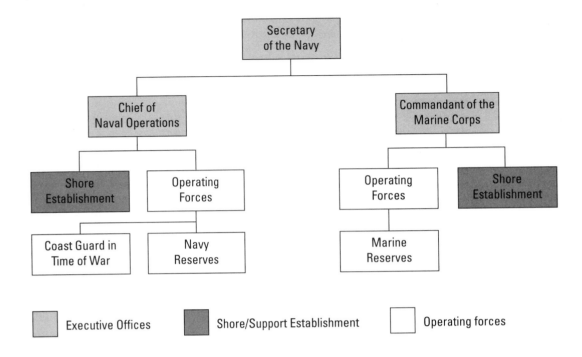

Navy shore and operating force commands are assigned to an echelon level. The echelon level serves to identify each command's Navy administrative chain of command. There are six echelon levels. The CNO is echelon 1, commanders that administratively report directly to the CNO are assigned to echelon 2, and commanders that administratively report to echelon 2 commanders are assigned to echelon 3. This pattern repeats to echelon 6. Destroyers and frigates are typically echelon 6 level commands. Echelon level implies neither command authority nor seniority. Commands of identical composition may be in different echelons depending on their administrative chain of command.

The operating force commanders, i.e., numbered fleet commanders, have a dual chain of command. Operationally, they typically report to the NCC, which reports directly to the CCDR. Administratively, they report to the NCC, who is normally an echelon 2 commander and reports to the CNO, which is responsible to provide trained and equipped naval forces to the CCDR.

NUMBERED FLEETS

The Navy has major and numbered fleets. Major fleets are Navy components for CCDRs that have assigned Navy numbered fleet(s). Typically the Navy numbered fleet commander reports operationally and administratively to the major fleet commander, who reports operationally to the CCDR and administratively to the CNO. Commanders of major fleets operate at the operational level supporting CCDR strategic objectives.

JP1-02 defines a numbered fleet as "a major tactical unit of the Navy, immediately subordinate to a major fleet command and comprising various TFs, elements, groups, and units for the purpose of prosecuting specific naval operations." Numbered fleet commanders command tactical forces but routinely operate at the operational level as either JTF commanders or JFMCCs. In addition, numbered fleet commanders normally are the Navy force commanders (COMNAVFORs) for subunified commanders and JTF commanders.

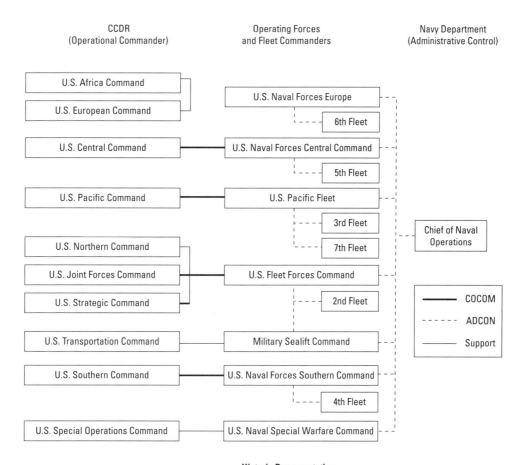

CCDR
(Operational Commander)

Operating Forces
and Fleet Commanders

Navy Department
(Administrative Control)

U.S. Africa Command

U.S. European Command

U.S. Naval Forces Europe

6th Fleet

U.S. Central Command

U.S. Naval Forces Central Command

5th Fleet

U.S. Pacific Command

U.S. Pacific Fleet

3rd Fleet

7th Fleet

Chief of Naval Operations

U.S. Northern Command

U.S. Joint Forces Command

U.S. Strategic Command

U.S. Fleet Forces Command

2nd Fleet

COCOM
ADCON
Support

U.S. Transportation Command

Military Sealift Command

U.S. Southern Command

U.S. Naval Forces Southern Command

4th Fleet

U.S. Special Operations Command

U.S. Naval Special Warfare Command

Historic Representation
Consult the current Secretary of Defense Global Force Management Implementation Guidance ((GFMIG) for current chain of command.

NAVY FORCES

Navy groups are composed of Navy forces. Navy forces are designed to be multimission and work above, on, and below the world's oceans. They are the basic building blocks of the Navy's tactical construct.

TASK ORGANIZATION

The Navy operational-level commander commands forces at the tactical level organized from individual platforms. Task organizing enables the operational-level commander to exercise a more reasonable span of control. Individual platforms are assigned and/or attached to a task force (TF). Each TF is assigned a commander, and only the commander reports to the operational commander. TFs also allow an operational commander to subdivide subordinate forces and delegate authority and responsibility to plan and execute based on mission, platform capability, geography, or other issues and challenges. The commander, task force (CTF) further subdivides the TF into task groups, units and elements to ensure span of control at the tactical level is maintained.

These subdivisions may be organized based on capabilities, missions, geography, or a hybrid of all three; they produce tactical and operational results to accomplish operational and strategic objectives to satisfy strategic goals toward fulfillment of national policy.

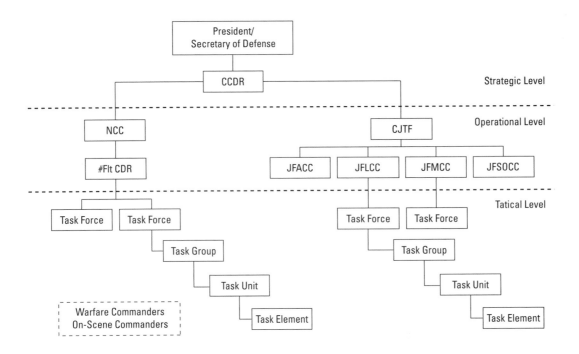

TASK ORGANIZATION NUMERICAL REPRESENTATION

A numerical representation is used to designate the command level of the task organization scheme.

The first numeral gives the TF its identity. Allied communication publication (ACP) 113 specifies the TF identities that countries can use for their forces. As of June 2008, the United States was allocated TF designations of 0–199, 201–299, and 800–824. Typically, task force identifiers for U.S. forces are designated with two or fewer digits. On occasion, the force component will utilize three digits to indicate a joint force assignment. Each task command is composed of forces from the higher task command and has a commander with command authority of OPCON, TACOM, TACON, or support. Only a commander with OPCON or TACOM can assign or attach units to subcomponents in the task organization structure.

Establishment of TFs ensures unity of command and unity of effort across the multimission platforms assigned for an operational mission or tasking. TF categories (Logistics, Navy Special Warfare, etc.) are established by doctrine.

Need. Is a TF needed? The purpose of task organizing is to produce tactical and operational results that enable accomplishment of operational and strategic objectives. What missions do the current operational effort require? Task forces execute discrete groups of missions that when coupled will accomplish a tactical or operational result. Can a discrete task or set of tasks that is not applicable to the entire maritime force be identified? Do the scale, scope, complexity, and duration of expected operations necessitate the creation of multiple TFs? Are there missions that all TFs will be required to execute? If yes, can the mission be executed by the establishing command or should a separate TF be created?

Mission. What operational and tactical effects will the missions expected of each TF accomplish? Are all missions expected to be performed by the maritime force assigned to TFs? Are the units assigned to TFs capable of accomplishing missions assigned to the TF? Who has the preponderance of forces necessary, the ability to characterize the operational environment, and the ability to command and control the forces assigned to achieve success?

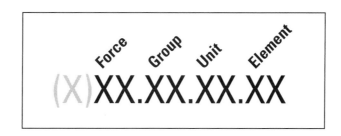

Span of Control/Command. There are limits as to how many subordinates a commander can effectively control. At some point, managing the incoming information will exceed the abilities of the commander and staff. Some studies suggest that humans are able to manage up to seven separate subordinate entities. However, given the fluid nature of events at the operational level, and the volume of information in today's operational environment, even not considering "the fog of war," it is even less likely that one commander can effectively oversee this number. Expanding beyond this threshold taxes the cognitive abilities and may cause a loss of proper focus or, in extreme instances, lead to organizational paralysis. One can reduce the number of subordinate commands by narrowing the span of control. This approach deepens the fighting organization by adding intermediate command echelons. Trade-offs exist between organizational width (number of TFs) and depth (fewer TFs, with each having multiple subordinate commanders) depending on the complexity of the operation, ability of the commander, staff capacities and expertise, capabilities of forces and communication networks, expanse of the operating area, the volume of information, and the skills of subordinate commanders and their staffs.

TF Relationships. What will be the relationship between TFs operating concurrently in the maritime AO? With multimission ships it is expected that ships in one TF have capabilities applicable to mission(s) assigned to another TF.

Flagship. TFs typically require a flagship or shore headquarters facility with sufficient communications and computer networks to effectively provide situational awareness (SA) to the operational commander, other TF commanders, and subordinate forces. Without this ability, the TF commander cannot effectively communicate, coordinate, and collaborate up, down, and across the force and may unintentionally disrupt the effective control of subordinate tactical forces.

MARINE CORPS ORGANIZATION

As with the Navy, Marine Corps operating force commanders have a dual chain of command. Operationally, they are the Service component and report to the appropriate CCDR. Administratively, they report to the MCC, which reports to the CMC, who is responsible to provide trained and equipped Marine Corps forces to the CCDR.

MARINE CORPS FORCES COMMANDS

Each CCDR has a maritime force (MARFOR) component commander assigned; however, only two have Marine forces assigned—Marine Corps Forces Command (MARFORCOM) and Marine Forces, Pacific (MARFORPAC) serve as the Service component commanders for U.S. Joint Forces Command (JFCOM) and U.S. Pacific Command (PACOM), respectively. MARFOR commanders are part of the Service or administrative chain of command and are responsible to the CMC for equipping, training, administering, and sustaining their forces. These forces include:

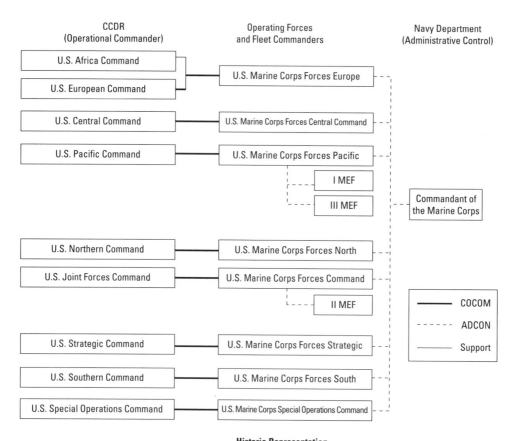

Historic Representation
Consult the current Secretary of Defense Global Force Management Implementation Guidance ((GFMIG) for current chain of command.

- Marine expeditionary force (MEF) based in southern California and Arizona assigned to MARFORPAC
- Marine Expeditionary Force based in North and South Carolina assigned to MARFORCOM
- Marine Expeditionary Force based in Okinawa, mainland Japan, and Hawaii assigned to MARFORPAC.

The Marine Corps Reserve is closely integrated with the active duty Marine Corps Forces. Within the Service chain of command, the commander of Marine Corps Forces Reserve provides Selected Marine Corps Reserve units and individual Marines to active duty Marine Forces when directed by the President/SecDef. When activated, reserve units are assigned to USJFCOM and administratively to MARFORCOM.

ORGANIZING FOR COMMAND AND CONTROL
Command and Control is the exercise of authority and direction by a properly designated commander. As a system, naval command and control has three components—command and control organization, information, and command and control support. *Command and control organization* encompasses the commander and the chain of command that connects superior commanders with subordinate commanders. *Information* is the lifeblood of the entire command and control system. *Command and Control support* is the structure by which the naval commander exercises command and control. It includes the people, equipment, and facilities that provide information to commanders and subordinates.

INDIAN OCEAN (September 4, 2008) Operations Specialist 3rd Class Sarah M. Hernandez, stands the global command and control systems–maritime (GCCS-M) watch in tactical flag command center (TFCC) aboard the Nimitz-class aircraft carrier USS *Abraham Lincoln* (CVN 72). *Lincoln* and embarked Carrier Air Wing (CVW) 2. (U.S. Navy photo by Mass Communication Specialist 3rd Class Ronald A. Dallatorre/Released)

THE COMPOSITE WARFARE COMMANDER (CWC) CONCEPT

The composite warfare commander (CWC) doctrine was developed initially in the late 1970s to provide Navy-wide standard procedures for command and control (C2) afloat. Its genesis was in response to the rapid growth in potential air and surface threats facing our naval forces during the Cold War. Realignment of surveillance and reaction responsibilities with continued emphasis on decentralized authority has led to more effective management of maritime resources for tactical sea control, thereby providing the officer in tactical command (OTC) greater opportunity to concentrate on the forces' primary mission of power projection. CWC doctrine was an effective C2 structure for open ocean operations against the global threat presented by the Soviet Union during the Cold War. The focus of the doctrine was defensive because the need was for a decentralized C2 methodology that could defend successfully against a large, capable threat at sea.

Notwithstanding the demise of the Soviet Union, potential air, surface, subsurface, and littoral threats facing our navy forces have continued to grow in recent years. These new threats have resulted from improved weapons, sensors, and delivery systems, and the proliferation of modern platforms employing these systems. Additionally, the overall strategic situation has changed in that now the primary threats may be regional rather than global, and operations often are in littoral instead of open-ocean areas.

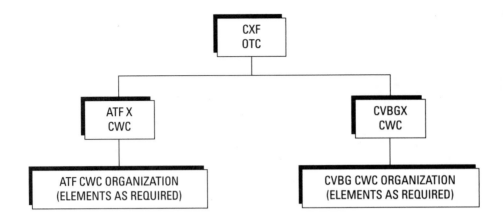

In response to these changes the emphasis has shifted from a C2 structure with a defensive focus to a methodology supportive of both offensive and defensive mission objectives.

The OTC will always be responsible for accomplishing the mission of the forces assigned. He may delegate authority for the execution of various activities in some or all warfare areas to designated subordinate warfare commanders and/or coordinators.

TACOM—The authority delegated to a commander to assign tasks to forces under their command for the accomplishment of the mission assigned by higher authority. This term is used primarily in maritime operations. It is narrower in scope than operational command but includes the authority to delegate or retain tactical control (TACON). (MMOPS-1)

OTC—In maritime usage, the senior officer present eligible to assume command, or the officer to whom the senior officer has delegated TACOM. (Joint Publication (JP) 1-02)

In many aspects of maritime warfare, it is necessary to preplan the actions of a force to an assessed threat and to delegate some command functions to a subordinate. Once such functions are delegated, the subordinate is to take the required action without delay, always keeping the OTC informed of the situation. The OTC and/or delegating commander retains the power to negate any particular action.

A centralized command is the most direct way of allowing the OTC to make use of his experience and ability. However, circumstances and command capabilities can make delegation necessary. Some factors are:

- Mission
- Threat
- Need for quick action or reaction
- Necessity to carry out many actions in different places at the same time
- Practical inability of the OTC to exercise all functions because of excessive workload or the requirements of some actions for specific knowledge of capabilities
- Lack of appropriate display and communication capabilities
- Force size and composition.

Decentralization of command is accomplished by delegating control of specific warfare areas and function among principal warfare commanders, functional commanders, and coordinators. The difference between principal warfare commanders and functional commanders and supporting coordinators is important. When authorized by the CWC, warfare commanders have tactical control of resources assigned to them and may autonomously initiate action. Supporting coordinators execute policy, but do not initiate autonomous actions.

Officer in Tactical Command

Composite Warfare Commander

Principle Warfare Commanders

Air Defense Commander (ADC)	Antisubmarine Warfare Commander (ASWC)	Information Warfare Commander (IWC)	Sea Combat Commander (SCC)	Strike Warfare Commander (STWC)	Surface Warfare Commander (SUWC)

Functional Warfare Commanders

Maritime Interception Operations Commander (MIOC)	Mine Warfare Commander (MIWC)	Operational Deception Group Commander	Screen Commander (SC)	Underway Replenishment Group (URG) Commander

Coordinators

Air Resource Element Coordinator (AREC)	Airspace Control Authority (ACA)	Cryptologic Resources Coordinator (CRC)	Force Over-the-Horizon Track Coordinator (FOTC)	Force Track Coordinator (FTC)	Helicopter Element Coordinator (HEC)	Launch Area Coordinating Authority (SOCA)	Submarine Operations Coordinating Authority (SOCA)	TLAM Strike Coordinator (TSC)

Responsibilities

OFFICER IN TACTICAL COMMAND

The OTC has overall responsibility for accomplishing the mission of the force. The OTC's policy and procedure for succession of command authority as well as designation of the standby OTC, should be specified in advance of the operation in the OTC's orders. The OTC shall specify the chain of command between himself and, when designated, the CWC, PWCs, functional commanders, coordinators, and the force under their TACON.

COMPOSITE WARFARE COMMANDER

The commander to whom the OTC has delegated specific authority for force direction and control of warfare functions.

CWC—The OTC is normally the CWC. However, the CWC concept allows an OTC to delegate TACOM to the CWC. The CWC wages combat operations to counter threats to the force and to maintain tactical sea control with assets assigned; while the OTC retains close control of power projection and strategic sea control operations. (JP 1-02)

PRINCIPAL WARFARE COMMANDERS

Subordinate to the OTC and CWC are five PWCs: Air Defense Commander (ADC), Antisubmarine Commander (ASWC), Information Warfare Commander (IWC), Strike Warfare Commander (STWC), and Surface Warfare Commander (SUWC). Dependent on the situation, the ASWC and SUWC can be linked and put under an SCC. The warfare commanders are responsible for collecting and disseminating information and, in certain situations, are delegated authority to respond to threats with assigned assets.

FUNCTIONAL WARFARE COMMANDERS

The OTC or CWC may form temporary or permanent functional groups within the overall organization. Functional groups are subordinate to the OTC and CWC and are usually established to perform duties which are generally more limited in scope and duration than those acted upon by PWCs. Examples of functional commands include: MIOC, MIWC, Operational Deception Group Commander, Screen Commander (SC), and Underway Replenishment Group (URG) Commander. Functional Warfare Commanders (FWCs) are responsible for collecting and disseminating information and in certain situations are delegated authority to respond to threats with assigned assets.

COORDINATORS

Coordinators are asset and resource managers. They carry out the policies of the OTC or CWC, if assigned, and respond to the specific tasking of either principal or FWCs. Coordinators differ from warfare commanders in that coordinators execute policy but do not initiate autonomous actions nor do they normally exercise TACON over assigned assets. Examples of coordinators include: Air Resource Element Coordinator (AREC), Airspace Control Authority (ACA), Cryptologic Resource Coordinator (CRC), Force Over-the-Horizon Track Coordinator (FOTC), Force Track Coordinator (FTC), Helicopter Element Coordinator (HEC), Submarine Operations Coordinating Authority (SOCA), TLAM Launch Area Coordinator (LAC), and TLAM Strike Coordinator (TSC).

INFORMATION

Two Types of Information

Information is absolutely essential to effective Command and Control. There are two basic types of information.

Image-building information is used to help create an understanding of the situation in order to make a decision. For instance if the CWC is to assign assets properly to subordinate commanders, he or she must have a certain amount of information in order to make good assignments.

Execution information is used as a means of coordinating actions in the execution of a plan after a decision has been made. In this case the CWC has assigned assets and now the subordinate commanders must use new information to properly employ the assets.

QUALITY OF INFORMATION

There are six attributes used to describe the quality of information. Although by no means all-inclusive, these characteristics provide a basis for qualitative assessment that applies to the mission, task, or situation at hand:

Accuracy—Information that conveys the true situation.

Timeliness—Information that is available in time to make decisions.

Usability—Information that is in common, easily understood formats and displays.

Completeness—All necessary information required by the decision maker.

Precision—Information that has the required level of detail or granularity.

COMMAND AND CONTROL SUPPORT

Effective command and control support helps the naval commander unify the force in the face of disorder (Fog of War) and shape the course of events to achieve a specific goal. Further, it helps the commander function effectively across the full range of conflict, in any environment and helps generate a rapid tempo of operations, while coping effectively with disruptions created by the enemy. Moreover, although our philosophy of command and control is based on our warfighting needs, it applies equally to successful mission accomplishment during operations other than war.

COMBAT INFORMATION CENTER (CIC)

The principal shipboard facility for command and control support is called Combat Information Center (CIC). Modern warfare requires the handling of vast amounts of information and fast reaction. To this end, CIC acts as the ship's tactical "nerve center" where most major decisions in fighting are made by the Commanding Officer or, in his stead, the Tactical Action Officer (TAO). CIC is the place where the requirements and orders of the CWC and warfare commanders are translated into action for that individual ship.

MISSION OF CIC

The mission of a ship's Combat Direction Center is to *provide command and control stations with tactical and strategic information correlated from all sources to enable the Commanding Officer to determine the proper course of action in a multi-threat environment.*

Primary Function of CIC

The primary function of CIC is *information control and handling, which involves collecting, processing, disseminating, and protecting pertinent tactical information.*

- *Collecting:* Gathering and formatting data for processing.
- *Processing:* Filtering, correlating, fusing, evaluating, and displaying data to produce image-building information required for commanders to take appropriate action.
- *Disseminating:* Distributing image-building or execution information to appropriate locations for further processing or use.
- *Protecting:* Guarding our information from an adversary's attempts to exploit, corrupt, or destroy it.

SECONDARY FUNCTIONS OF CIC

The secondary functions of CIC are to support and assist.

- *Support functions*
  - Radar reporting
  - Communications control
  - Aircraft control
  - Control of small boats and landing craft
  - AW/AuSW/SUW
  - NSFS

- *Assist functions*
  - Surface and air contact tracking and reporting: This includes computing contact course and speed, identification, and recommendations for maneuvering in accordance with the Rules of the Nautical Road for surface contacts.

  - Maintaining a current navigational plot for ocean transits: CIC's DR plot is updated from the Navigator's electronic and celestial fixes.

  - Search and Rescue: CIC coordinates all information received and provides the OOD with course recommendations, communications, and coordination with other units involved in any search and rescue effort.

  - Tactical maneuvering: CIC provides a back-up for signal coding/decoding, recommendations to the OOD for proper maneuvering, communications log of all tactical signals.

  - Low Visibility Piloting: CIC directs the movements of the ship based on an accurate radar navigation plot during periods of rain, fog, snow, etc. The OOD is still responsible for the safety of the ship and must balance CIC's recommendations with what he can see and hear on the bridge.

CIC ORGANIZATION

CIC organization involves the assignment and utilization of personnel to accomplish specialized evolutions consistent with information handling and control/assist functions. Personnel billets are assigned by the Ship's Manning Document, and watch stations are assigned by the Watch Quarter and Station Bill, which provides details on each individual's watch assignments and responsibilities. The CIC watch organization fluctuates, along with the rest of the ship's watch organization, in accordance with the following readiness conditions:

Condition I—Combat posture in the multi-threat environment (condition of maximum readiness, GQ).

Condition II—Modified general quarters—used to permit some relaxation among the crew during combat readiness conditions. Condition II may also be used to set battle stations for a specific threat (e.g. IIAS is set when there is only a submarine threat).

Condition III—Wartime cruising. Extended Defensive Profile (man all major sensor and weapons systems with less than maximum readiness, i.e., one gun mount manned out of two). This allows for rapid response in a lower threat environment.

Condition IV—Peacetime cruising. Ensure safe navigation.

CIC MANNING

The following are major stations which are manned in CIC during Condition IV:

CIC Watch Officer—Coordinates the entire CIC team and ensures information is properly evaluated and disseminated.

CIC Watch Supervisor—A senior enlisted petty officer supervising all information and watch station performance.

Radar/NTDS Operators—Control radar repeaters and report contact range/bearing/altitude information.

Plotters—Maintain current displays or plots on air, surface (maneuvering board), geographic (DRT), strategic, and formation plots.

Status Board Keepers—Maintain current displays of data collected and ensure this data is continually updated.

R/T and S/P Phone Talkers—Receive and transmit information as directed.
Navigation Plotters—Maintain navigation plot on same scale chart as bridge using electronic measures: radar, GPS, LORAN, DR.

Conditions I, II, and III require additional manpower as weapons consoles are manned in order to defend against a possible threat. Manned weapons systems require a greater level of supervision. The primary supervisor in Combat during Conditions I, II, and III is the Tactical Action Officer (TAO), the Commanding Officer's direct representative in CIC tasked with "fighting the ship."

Naval Tactical Data System (NTDS)

The Navy Tactical Data System (NTDS) is a critical part of effective command and control support systems, allowing for real-time information flow between all platforms having NTDS capabilities. The system transmits target position, designation (track number), and weapons assignment status to all other NTDS platforms. This type of information link allows earlier detection (because you are not limited to the range of your own sensors), thus allowing more time for the CWC, Warfare Commanders, COs, and TAOs to make a sound tactical decision in the minimum time. Also, by indicating whether another platform has engaged a target, it helps prevent a duplication of effort, freeing other weapons systems for other targets. *NTDS is one of the key functional capabilities which allows the Composite Warfare Concept to work.* It allows automatic transmission of information from one ship or aircraft to another without human interface. The common radar picture allows every NTDS capable platform to see what every other NTDS platform sees. NTDS is a high speed time sharing system controlled by the NTDS computer on one unit known as the NET CONTROL SHIP (NCS). The NCS interrogates each unit (known as a picket) sequentially and that unit sends its data to all others. The net cycle time, or time it takes to interrogate all Participating Units (PU) depends on the number of PUs, but in no case will exceed 4 seconds. If a unit fails to respond to an interrogation because of communications or computer failure, the next ship is then interrogated and the missed unit will not be interrogated again until the next cycle. Should the NCS unit's computer fail, a predesignated alternate will switch into the NCS mode and assume duties as NCS. There can only be one NCS at any given time.

TRACK NUMBERS

All NTDS symbols are assigned track numbers within the system to allow for a common method of tracking the status. Track numbers are octal numbers assigned by the computer (0000–8888) to each NTDS symbol.

SYMBOLOGY

NTDS symbology allows every ship to classify a contact as hostile, friendly, or unknown. Further subclassification as surface, subsurface, or air aids the CWC, Warfare Commanders, CO, and TAO in determining at a glance the tactical situation.

SPEED LEADERS

Each moving symbol will have a line coming from the origin of the symbol indicating (the relative direction speed of movement.) Speed leaders of air contacts indicate the distance traversed in one (1) minute. The magnitude of surface and subsurface contacts shows the distance traveled in three (3) minutes. On fleet consoles, there is an option allowing you to adjust speed leaders to reflect movement for any period of time up to 30 minutes. This is used mostly to time air intercepts.

DATA LINKS

1. Link 11-(TADIL A)-Two-way link between NTDS link-capable units (could be surface, air, or sub). Data is displayed on NTDS consoles.
2. LINK 4A-(TADIL C)-Two-way data link between NTDS ship and interceptor aircraft. Along with data exchange, engagement and intercept orders may be transmitted to the aircraft via Link 4A.
3. LINK 16-(TADIL I)-Identical in purpose to LINK 11 and LINK 4A; however, it provides significant improvements such as nodelessness, jam resistance, flexibility of communication operations, separate transmission and data security, increased numbers of participants, increased data capacity, and secure voice capability.

REVIEW QUESTIONS

1. Diagram the organization of the Department of the Navy.
2. Describe the administrative and operational Chains of Command for the Navy.
3. Describe the administrative and operational Chains of Command for the Marine Corps.
4. List the Major and Numbered Fleets and their relationships.
5. What are the fours steps for processing and handling information?
6. List three support functions provided by CIC.
7. How does CIC support the Officer of the Deck?
8. Which Condition of Readiness is set at General Quarters?
9. What are the duties and responsibilities of the CIC Watch Supervisor?
10. What is Operational Control (OPCON)?
11. Who exercises Tactical Control (TACON)?
12. What is the concept of leadership employed by the CWC?
13. What is image-building information used for?

CHAPTER 5

Operational Design

LEARNING OBJECTIVES

At the end of this chapter the student will be able to:

- Understand the requirements for operational Operational.
- Discuss the process of operational design.
- Explain both center of gravity and critical vulnerability.
- Describe commander's intent.
- Define the key steps during the mission analysis phase of the planning process.
- Define specified, implied, and essential tasks.

Introduction

The Navy leaders of World War II were practitioners of operational art and design long before these terms were adopted in joint doctrine. While technology has provided today's Navy leaders with a vast array of capabilities, the fundamental underpinnings of operational planning and operational decision making have not changed since Halsey and Spruance were fleet commanders. The chapter shall then show how operational art, operational design, and operational planning fold into today's Navy operational decision-making processes, which have evolved from the ideas and concepts that were learned on the high seas and littorals during World War II.

The admirals believed that the planning process developed prior to the war had become overly complicated and too mechanical. The results of this thinking are articulated clearly in the 1948 edition of the Naval Manual of Operational Planning, which Admiral Spruance developed for the CNO. The manual began very simply:

> In many ways the solution of a military problem is similar to the solution of the problems of everyday life. When one is faced with a problem, he studies the situation from all angles and decides what must be done. He then determines how it should be done—and proceeds to act, governing his actions according to the development of the situation. Hundreds of such problems are solved by everyone. Great majorities of them are not committed to writing, nor are they consciously solved in phases. They are solved by the natural, logical thought process normally pursued by the human mind.

In military problems the situation is much more complex, and the stakes are much higher. In order to ensure a logical thought process, to guard against the oversight of important details, and to form a readily available record, military problems are usually solved in written form.

In war there is no second prize; there is seldom a second chance.

Achievement of objectives does not lend itself to mechanistic, deterministic, scientific models or simple linear processes—developing a solution requires study of the interplay of literally hundreds, if not thousands, of independent variables. In other words, developing a solution for strategic objectives is more of an art than a science. Operational art serves as a bridge and as an interface between maritime strategy and naval tactics. It is the application of creative imagination by commanders and staffs—supported by their skill, knowledge, and experience—to design strategies, campaigns, and major operations and to organize and employ military forces. Operational art is the thought process commanders use to visualize how best to efficiently and effectively employ military capabilities to accomplish their mission.

Operational art helps the operational commanders overcome the ambiguity and uncertainty of a complex operational environment. It governs the deployment of forces, their commitment to or withdrawal from a joint operation, and the arrangement of battles and major operations to achieve operational and strategic military objectives. Among the many considerations, it requires commanders to answer the following questions:

- What conditions are required to achieve the objectives? (Ends)
- What sequence of actions is most likely to create those conditions? (Ways)
- What resources are required to accomplish that sequence of actions? (Means)
- What is the likely cost or risk in performing that sequence of actions? (Risk)

Operational Design

Operational design14 provides a deep understanding of the problem to be solved and sets the conditions for tactical success. Operational planning—particularly for extensive operations that require a campaign—uses various elements of operational design to help commanders and staffs visualize the arrangement of force capabilities in time, space, and purpose to accomplish the mission. Operational design is the conception and construction of the framework that underpins an operational plan and its subsequent execution. While operational art is the manifestation of informed vision and creativity, operational design is the practical extension of the creative process. Together they synthesize the intuition and creativity of the commander with the analytical and logical process of design. Operational design must include:

- Understanding the strategic guidance
- Identifying the enemy's critical strengths and weaknesses
- Developing an operational concept that will achieve strategic and operational objectives

These three key considerations become the framework for planning a campaign or major operation, identifying the enemy's centers of gravity and critical factors, and developing an operational concept to achieve strategic objectives.

During execution, commanders and planners continue to consider design elements and adjust current operations and future plans to capitalize on tactical and operational successes as the operation unfolds.

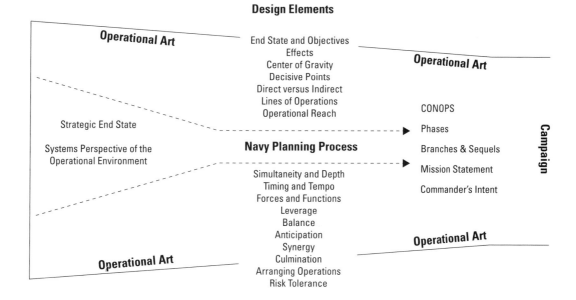

Throughout these efforts, the following five principles are highlighted as keys for successful operational design:

- Determine what needs to be done and why, but not the tactical specifics of how.
- Organize subordinate commanders to take best advantage of all of the military force capabilities.
- Articulate the geometry of the operational environment to provide sufficient control measures in terms of boundaries and fire control measures without over controlling the fight.
- Establish command relationships that promote interdependence among the components, instill a "one team, one fight" mentality, provide command authority commensurate with responsibilities, and build trust and confidence.
- Decentralize authorities to empower subordinates to operate within the commander's intent and take advantage of unforeseen opportunities.

Elements of Operational Design

Operational art encompasses operational design—the process of developing the intellectual framework that will underpin all plans and their subsequent execution. The elements of operational design are tools to help support operational commanders and their staffs visualize what the campaign should look like and to shape the commander's intent.

The operational environment is not the only factor that affects operational design. Other factors, such as the availability of host-nation support (HNS), diplomatic permission to overfly nations and access en route air bases, the allocation of strategic mobility assets, the state of the operations area infrastructure, and forces and resources made available for planning all have an impact on the operational design. In the final analysis, the goals of a sound operational design are to ensure a clear focus on the ultimate strategic objective and corresponding strategic centers of gravity (COGs), and provide for sound prioritization, sequencing, synchronization, and integration of all available military and nonmilitary instruments of power to that end.

The commander uses the tenets of operational art and design to define the mission (what the commander has been told to do and the reason for it) and assemble/examine information relating to the mission. This information constitutes the initial estimate of the situation. Also critical for follow-on plan development, the initial estimate of the situation provides a loose collection of diverse references that the operational commander and staff can consider in the development of the basic plan and continuous estimate-of-the-situation refinement. Information collected from exercising operational art and design is cataloged as follows:

- Determination of specified, implied, and essential tasks
- Ultimate and intermediate objectives
- Force requirements
- Balancing of operational factors against the objectives
- Identification of the enemy and friendly operational centers of gravity, critical vulnerabilities, and decisive points (DPs)
- Initial lines of operations (LOOs)
- Direction (axis)
- The operational idea
- Operational sustainment

DETERMINING SPECIFIED, IMPLIED, AND ESSENTIAL TASKS

Every mission consists of two elements: the tasks to be accomplished by one's own forces and the purpose of those tasks. Before going further, it is necessary to illustrate how tasks, operations, and missions are related. If a mission or operation has multiple tasks, then the priority of each task should be clearly expressed.

Specified tasks are specifically assigned to a unit by higher headquarters. Specified tasks are derived primarily from the execution paragraphs of the operation order (OPORD), but they may be found elsewhere, such as in the mission statement, coordinating instructions, or annexes. Implied tasks are not specifically stated in the higher headquarters order but must be performed in order to accomplish specified tasks. Implied tasks emerge from analysis of the order, the commander's guidance, and the enemy. Routine, inherent, or SOP tasks are not included in the list of tasks.

Those tasks that most contribute to mission success are deemed essential, and they become the central focus for operations planning. Essential tasks are those that define mission success and apply to the force as a whole. Essential tasks can come from either specified or implied tasks. If a task must be successfully completed for the commander to accomplish his purpose, it is an essential task. Only essential tasks are included in the proposed mission statement. The following example shows the three types of tasks that a commander may experience.

Though not elaborated in this example, the planning team also must determine the follow-on tasks that may be required at a later time due to the impacts of the operation, the situation in the operational environment, the enemy's actions, and the dynamic nature of the operational environment. These tasks, commonly seen in a directive or guidance as "be prepared to" or "BPT," shape the planning team's efforts as well as the specified, implied, and essential tasks.

Determination of enemy objectives, enemy courses of actions (COA), critical vulnerabilities, and decisive points is completed as part of the intelligence preparation of the operational environment (IPOE) process before friendly mission analysis and COA development are conducted.

If the COG is a military force, one may find the operational functions, as detailed in joint doctrine, a useful starting point for the discovery of a force's critical capabilities:

- Command organization
- Fires

- Intelligence
- Movement and maneuver
- Protection
- Sustainment

While these operational functions are a useful starting point for initiating analysis of critical capabilities, a planning staff must also broaden its examination to include other tangible or intangible factors that might be critical capabilities for a COG. In addition, if the enemy and its corresponding objective are found in a less conventional environment, such as an insurgency, the COG and related critical capabilities will likely reflect more intangibles, such as the support of the local populace, information control, and others of such ilk.

Each critical capability (CC) has a number of associated critical requirements. For example, a CC of sustainment has critical requirements for bases, logistics ships, fuel, ammunition, food, etc. A critical requirement for operational protection may lead to critical requirements for screening ships, combat air patrol aircraft, computer network protection, operations security (OPSEC), land-based patrol aircraft, a submarine devoted to protecting the center of gravity, etc. A critical requirement for operational fires may lead to the requirement for long-range strike aircraft, long-range strike missiles, special operations forces, etc. The CC for operational command organizations leads to critical requirements for an understanding of command relationships and functions, for the hardware, software, and networks that support the commander's ability to receive and send information. Deficiencies in critical requirements are critical weaknesses. These critical weaknesses, if closely related to a COG, may prove to be valuable paths for one to attack a COG. Critical weaknesses that can be exploited by an opposing force are critical vulnerabilities.

THE OPERATIONAL IDEA

The operational idea is the very core of a design for a major operation. It is the framework upon which a CONOPS is built. Main features of an operational idea are:

- Full use of deception
- Speed of execution
- Multiple options
- No discernible patterns
- Creative and novel

Operational ideas are normally captured graphically using operational vice tactical symbols and:

- Describe in broad terms what friendly forces have to do, where, and when.
- Express in broad and succinct terms the type of actions by all forces in the AO.
- Present the sequence of accomplishing a given strategic or operational objective.

OPERATIONAL SUSTAINMENT

How forces will be sustained and how the military effort will be maintained is an integral part of operational design. Though sustainment is related to the operational function of sustainment, sustainment carries a greater degree of emphasis in the design phase because of time. Operational sustainment applies from the very beginning of an operation until the objective is achieved; furthermore, sustainment applies to providing support for the forces even after the objective has been achieved. The need for sustainment in a major operation cannot be wished away.

DECISIVE POINTS

Identification of decisive points (DPs) remains an important feature of the COG analysis and its subsequent defeat or neutralization. Joint doctrine defines decisive point as "a geographic place, specific key event, critical factor, or function that, when acted upon, allows commanders to gain a marked advantage over an adversary or contribute materially to achieving success." The value of a DP is directly related to its relationship to a COG and its objective. A DP is neutral in nature.

DEVELOPMENT OF THE PLAN

The operational planning process is conducted continuously in peacetime and in conflict or war. It encompasses the entire range of military operations in which U.S. military forces could be employed to accomplish national or multinational strategy objectives.

The NPP assists commanders and their staffs in analyzing the operational environment and distilling planning information in order to provide the commander a coherent framework to support decisions. The process is thorough and helps apply clarity, sound judgment, logic, and professional expertise. The NPP organizes these procedures into six steps, which provide commanders and their staffs a means to organize planning activities, transmit plans to tactical forces, and share a critical common understanding of the mission.

THE DIRECTIVE

The directive is communication from the operational commander that starts or governs action. In issuing a directive the commander has certain definite responsibilities; these are:

- To ensure that subordinates understand the situation by giving them pertinent available information.
- To set forth clearly the task to be achieved by the entire force, as well as the task to be accomplished by each primary subdivision of that force.
- To provide each TF with adequate means to accomplish the assigned task.
- To allow subordinates appropriate discretion within the limits of their assigned spheres of actions. Necessary coordination is never sacrificed to accomplish this. The personality and ability of each subordinate are a consideration in determining the degree of discretion that can be entrusted.

The directive itself will be more easily understood, and will better convey the will and intent of the commander, if it is clear, brief, and positive.

- Clarity demands the use of precise expressions. A precise expression can have only one interpretation.
- Brevity calls for the omission of unnecessary and unnecessarily detailed instructions. Terse sentences are preferable when they convey the meaning clearly. However, clarity is never sacrificed for brevity.
- Positiveness of expression suggests the senior's fixity of purpose, with consequent inspiration to subordinates to prosecute their task with determination. The use of indefinite expressions leads to suspicion of vacillation and indecision. They tend to impose upon subordinates responsibilities fully accepted by a resolute commander.

Experience has shown that military directives are usually most effective if cast in a standard form well known to originator and recipient alike. Such a form tends to ensure against the omission of relevant features and minimizes possibilities of error and chances of misunderstanding. Operational commanders typically communicate direction using either a concept plan (CONPLAN), OPLAN, or operation order (OPORD).

These directives most frequently use the standard five-paragraph format briefly described below. The five basic paragraphs for all plans and orders are:

Paragraph 1:   Situation. This paragraph, the commander's summary of the general situation, ensures that subordinates understand the background for planned operations. It often contains subparagraphs describing enemy forces, friendly forces, and task organization, as well as higher headquarter's guidance.

Paragraph 2:   Mission. The commander inserts his own restated mission developed during mission analysis. This is derived from the mission analysis step and contains those tasks deemed essential to accomplish the mission.

Paragraph 3:   Execution. This paragraph expresses the commander's intent for the operation, enabling subordinate commanders to better exercise initiative while keeping their actions aligned with the operation's overall purpose. It also specifies the objectives, tasks, and assignments for subordinate commanders. It should articulate not only the objective or task to be accomplished but also its purpose, so that subordinate commanders understand how their tasks and objectives contribute to the overall CONOPS. The execution paragraph can be written in phases to convey the flow of the operation in an easy-to-understand, more logical progression.

Paragraph 4:   Administration and Logistics. This paragraph describes the concepts of support, logistics, personnel, public affairs, civil affairs, and medical services. The paragraph also addresses the levels of supply as they apply to the operation.

Paragraph 5:   Command and Control (C2). This paragraph specifies command relationships, succession of command, and the overall plan for communications and control.

REVIEW QUESTIONS

1. Operational art requires commanders to answer what questions?
2. The logical process of operational design must include.
3. Explain both center of gravity and critical vulnerability.
4. Describe commander's intent.
5. Define the key steps during the mission analysis phase of the planning process.
6. Define specified, implied and essential tasks.

CHAPTER 6

The Objective of Naval Warfare

LEARNING OBJECTIVES

At the end of this chapter the student will be able to:

- Demonstrate a basic understanding of sea control and power projection.
- List the goals of U.S. naval strategy.
- Describe both major and supporting warfare tasks.
- Give some examples of the use of naval forces in modern conflict.

ADDITIONAL READING

Naval Doctrine Publication 1, *Naval Doctrine*. CNO, Washington D.C., 19 May 1996.

Introduction

Strategy applies in peace as well as war. Activities at the strategic level focus directly on national policy objectives. National strategy coordinates and focuses all components of national power toward an objective. Military strategy deals with the application of force to secure the objective.

National and military strategy are determined by the President and Congress and promulgated annually by the Secretary of Defense. They determine naval strategy and thus the size and composition of naval forces. See appropriate NDPs for discussion of naval strategy as it generates force requirements and planning, employment, and readiness doctrine.

The goals of U.S. naval strategy are to:

- Deter aggression through a visible, forward deployed presence.
- Support and cooperate with allies in coalition warfare against aggressors.
- Conduct offensive missions in forward positions if deterrence fails.
- Conduct defensive missions in support of offensive operations.
- Defend the homeland and its seaward approaches.

Naval Strategy

Naval strategy involves the oceans, littoral areas, and contiguous land. Along with naval forces, strategists consider other national assets involved in the use activities of oceans and seas, including resource access, the Merchant Marine, space-based assets, and joint and allied forces.

The littoral comprises two segments of battle space:

- Seaward: the area from the open ocean to the shore which must be controlled to support operations ashore.
- Landward: the area inland from the shore that can be supported and defended directly from the sea.

Characteristics of the littoral region usually include:

- Confined and congested water and airspace occupied by a mix of friends, adversaries, and neutrals. Identification and early warning may be difficult.
- Difficult physical and oceanographic environment, posing both technical and tactical challenges.
- Layered and possibly sophisticated defenses. Many regional powers possess sophisticated weaponry. Others do not. Thus, there is a wide range of potential challenges.

Mastery of the littoral should not be presumed. It does not derive directly from command of the high seas. It requires focused skills and resources.

Naval surface forces are ideal for control of the littoral. They are particularly well suited for forward presence and crisis response missions. They provide powerful yet unobtrusive presence, strategic deterrence, and extended, continuous, and flexible crisis response. They project precise power from the sea and provide sea control if larger scale warfighting is required.

Should their regional deterrence mission fail, the Navy and Marine Corps team, both active and reserve, will provide the initial "enabling" capability for joint operations in conflict, as well as continued participation in any sustained effort. We will be part of a "sea-air-land" team trained to respond immediately to the unified commanders as they execute national policy.

Naval Missions

In rough order of increasing violence, possible naval missions include presence, deterrence, sea control, power projection, and warfighting.

Surface ships are the essential core of all naval missions. They are the only type of platform that can fight in all warfare areas simultaneously while providing the C2 capability to coordinate supporting missions of naval aircraft and submarines.

If sea control remains unchallenged, the primary naval mission will probably be provision of expeditionary forces, operating forward, from the sea, and tailored to national needs.

In this mission, surface ships will provide the primary conventional weapons, and the volume to carry the large loads of ordnance required. They can also carry tactical nuclear weapons should that challenge arise in a future regional crisis.

In the sea control mission, surface ships provide the overt presence and combat systems required to defend against aircraft and other ships. They are also a key part of the air-surface-submarine team necessary to conduct ASW.

SEA CONTROL

Sea control is the basis of national use of the sea. It is a force's ability to control the subsurface, surface, and airspace of maritime areas. Sea control cannot be carried out without surface ships, primarily surface combatant. Surface ships control air superiority aircraft and employ anti-air missiles to control the airspace. Surface ships control surface areas. Surface ships are an important component of the air-surface-submarine team that conducts anti-submarine operations. Surface ships provide C2 platforms and volumes of ordnance to support all operations. Land-based aircraft can support sea denial operations. Submarines can deny the surface and subsurface, But to exercise true sea control, tactical groupings of surface combatants are necessary, assisted by, or in support of, aircraft and submarine forces. Sea control is a prerequisite for all other naval operations. If not possessed when operations begin, it must be gained by:

- Destruction of enemy forces
- Neutralization of enemy forces by deterring their employment
- Destruction or neutralization of the land-based systems and infrastructure that supports enemy sea control or sea denial forces (power projection in support of sea control)
- Barrier and mining operations that prevent enemy egress to the open sea
- Area antisubmarine operations to destroy enemy submarines before they reach the open sea
- Defense of local sea areas for special purposes (local sea control)

Sea control provides:

- Protection of U.S. sea lines of communication (SLOCs)
- Denial of enemy commercial and military use of the seas
- Establishment of an area of operations for conducting strike warfare against land and in support of expeditionary operations
- Protection of logistics lines to forward deployed battle forces

POWER PROJECTION

As a highly sustainable force on scene, a naval force commander can command a joint task force while an operation is primarily maritime and shift that command ashore at the discretion of the unified commander. Focusing on the littoral area, the Navy and Marine Corps can seize and defend an adversary's port, naval, or air base to allow entry of heavy Army or Air Force forces. Power projection is a means of supporting land or air campaigns using naval forces. It can take the following forms:

- Tactical aircraft strike
- Cruise missile strike
- Amphibious assault or raid (with associated supporting arms)
- Shore bombardment
- Special warfare operations
- Minelaying
- Blockade, quarantine, or interception operations
- Joint and/or combined operations support
- Sealift to ensure heavy joint forces can arrive and fight effectively in major crises

Power projection is often an essential part of sea control in that the seizure of islands, chokepoints, peninsulas, and coastal bases may be necessary. The tasks the support power projection can be subdivided into two groups Major warfare tasks and supporting warfare tasks.

Major Warfare Tasks

The major warfare tasks are:

- AAW—destruction or neutralization of enemy aircraft/missiles
- ASW—destruction or neutralization of enemy submarines
- Antisurface warfare (ASUW)—destruction or neutralization of enemy surface ships
- Strike warfare (STW)—strikes against land targets
- Amphibious warfare (AMW)—land attack over the beach from sea-based forces
- Mine warfare (MIW)—minelaying and mine countermeasures operations.
- Command and control warfare (C2W)—destruction, control, or neutralization of C2 targets through integrated employment of destructive and non-destructive electronic attack (EA), electronic protection (EP), electronic warfare support (ES), surveillance, and command, control, communications, computers, and intelligence (C4I) systems

Supporting Warfare Tasks

Supporting warfare tasks in both sea control and power projection are:

- Naval special warfare—employing nonconventional warfare
- Ocean surveillance—detecting, locating, and classifying ocean traffic
- C4I
- Counter C4I and electronic combat (EC)—employing electronics warfare, including deception
- Logistics—resupplying combat consumables

Joint Operations

Naval forces must be prepared to operate as part of a joint force, either as dominant component with others in support, or in support of a ground component commander, an air component commander, or special operations forces commander. Joint operations may employ all elements of naval warfare.

Naval forces prepare in particular for two missions:

- To secure and maintain sea control whenever and wherever directed.
- To conduct power projection missions as directed. In power projection, Navy/Marine forces can either conduct limited operations or establish a littoral lodgement through which heavy Army and Air Force units pass for major combat operations.

Naval and Joint Task Organizations

The commanders-in-chief (CINCs) are combatant commanders (COCOMs) who carry out National Command Authority taskings. During operations and exercises, a naval component commander (CINCLANTFLT, CINCPACFLT, CINCUSNAVEUR, USCINCCENT) directs naval forces, usually through the numbered fleet commanders. The numbered fleet commanders establish task forces, task groups, task units, and task elements sized and composed to accomplish specific missions.

A CINC may also directly establish an organization to accomplish a specific mission. In that case, he designates a joint force commander (JFC) or a commander, joint task force (CJTF), and assigns forces to achieve mission objectives. Again, the joint task organization is composed of task forces, task groups, task elements, and task units. Naval units may have several task designations, but may only be under the operational control of a single commander.

Functional Organization

At sea, naval forces operate as directed by the officer in tactical command (OTC) under the composite warfare commander (CWC).

If an amphibious objective area (AOA) is established, naval forces within the AOA operate as directed by the commander, amphibious task force (CATF), and when ashore, as directed by the commander, landing force (CLF), subject to the overall authority of CATF.

Naval functional organizations are frequently the forward element of a CINC's joint force because of their inherent C2, firepower, and sustainability. Traditional naval task structures are often preferred by the CINCs to meet operational commitments. These are the CSG with its embarked air wing and the ARG with its embarked Marine air-ground task force (MAGTF) (usually a Marine expeditionary unit (MEU)). Reality, in the form of asset availability and other constraints, frequently requires more tailored forces. Adaptive force packages are Navy and Marine Corps forces tailored to complete missions at levels requiring less combat power than that provided by a CVBG or ARG. Common variations of adaptive force packages are the MAG, special purpose MAGTF (SPMAGTF), sea control ARG, and direct support to special operations group.

Modern Conflict

The worst case uses of naval forces are global conventional or nuclear war. The most likely uses, however, are below that level. Tactical differences between the worst and most likely cases are:

- Attrition rates of surface combatants in worst case missions would probably be high.
- The political sensitivity of the likeliest worst case scenarios will be very high, with concomitant requirements for high levels of communication up the chain of command and sensitive ROE.

Naval force tactical employment also addresses the tactical differences among:

- Operations against a major power within the envelope of its land-based air (primary strike operations involving securing initial sea and air control) and with an ASW threat.
- Operations against a major power outside the envelope of its land-based air (primarily sea control operations) and with an ASW threat, though probably a lesser one.

Operations against a regional threat country which:

- Lacks a global maritime reach, regardless of its ground capability or political and economic influence.
- Can challenge opposing naval forces for significant areas of sea control within its region, including interdiction of SLOCs.
- Can project naval air and amphibious power within its region, with a significant offensive capability against its neighbors.
- Has weapon and sensor technology and warfare capabilities equal to or greater than its neighbors, and in some warfare areas may approach those of major powers.

Operations against a territorial threat country which:

- Lacks maritime interests and seafaring and maritime aviation experience. Can only threaten U.S. naval forces if they approach within land-based strike range of its coast.
- May possess a significant offensive strike capability against its neighbors.
- May possess an initially credible but not particularly robust air defense system (perhaps better than that of a regional threat country because of its smaller extent).
- May pose a credible anti-landing threat and achieve, at best, sea denial of limited areas, primarily within 50 nm of its coast. Its ability to defend its littoral against sustained power projection strikes by aircraft or cruise missiles is poor. Mines are a credible threat along these coastlines.

Cases for employment of naval forces can therefore be defined as follows:

- Global conventional or nuclear war within a major power's land-based air envelope. Surface combatant attrition is high. Careful consideration of gain/cost ratio of objectives is necessary.
- Global conventional or nuclear war outside a major power's land-based air envelope. Surface combatants usable with moderate attrition rates.

Regional conflict against a regional threat country within a major power's land-based air envelope (may involve direct participation by the major power, or threatened use of its forces in support of a regional threat country):

- Surface combatants usable with moderate attrition rates
- Tactics sensitive to escalation
- Tactics sensitive to ROE

Regional conflict against a regional threat country outside a major power's land-based air envelope:

- Surface combatants highly usable with low attrition rates
- Tactics relatively insensitive to escalation
- Tactics sensitive to ROE

Regional conflict against a territorial threat country within a major power's land-based air envelope (may involve direct participation by the major power, or threatened use of its forces in support of a regional threat country):

- Surface combatants usable with moderate attrition rates
- Tactics sensitive to escalation
- Tactics sensitive to ROE

Regional conflict against a territorial threat country outside a major power's land-based air envelope:

- Surface combatants highly usable with low attrition rates
- Tactics insensitive to escalation
- Tactics highly sensitive to ROE

REVIEW QUESTIONS

1. What are the cases for the employment of naval forces?
2. What are the uses for naval forces in modern conflict?
3. Give some examples of major and supporting warfare tasks.
4. What is necessary for sea control and what does it provide?
5. What is a littoral region?

CHAPTER 7

Logistics

LEARNING OBJECTIVES

At the end of this chapter the student will be able to:

- Describe the seven principles of logistics.
- Explain how the levels of war and logistical support interact.
- Describe the different areas required for effective logistical support.
- Explain the supply planning processes.

ADDITIONAL READING

Conrad, Scott W. *Moving the Force: Desert Storm and Beyond* (Washington, D.C.: National Defense University, December 1994)

Dyer, George C. *Naval Logistics* (Annapolis, M.D: Naval Institute Press, 1960).

Wildenberg, Thomas *Gray Steel and Black Oil* (Annapolis, M.D: Naval Institute Press, 1996).

Navy Warfare Publication 4-09. Other Logistics Services, Parts I–VIII

Introduction

Effective naval logistics enables us to carry out the Navy and Marine Corps' assigned roles. It supports our ability to conduct continuous forward presence, peacetime engagement, deterrence operations, and timely crisis response from the challenging maritime and littoral environment. Through our logistics systems, Navy and Marine Corps striking power is always available, and always sustainable through established support systems.

Sustained forward deployment of naval forces also allows our nation to pursue regional coalition-building and collective security efforts. Thus, naval logistics forces must be able to provide and receive support within a variety of organizational structures. Consequently, engagement in joint and multinational logistics efforts are increasingly critical to support mutual readiness and capability, enhancing the efficiency and effectiveness of our combat operations.

Naval logistics operations are conducted much the same in peace as they are in war. They support and sustain the warfighter whenever and wherever, differing mainly in the magnitude of the requirements placed on the logistics systems and the level and types of threat to which these systems are exposed. A viable, accessible, and ready reserve of trained personnel and effective equipment, and reliable sources of war materiel, must back active logistics forces. These resources must also include agreements and understandings that permit the sharing of logistics resources among other services, other nations, and the private sector of all engaged nations.

Principles Of Logistics

Naval logistics—provided at the strategic, operational, and tactical levels; organized within the six major functional areas; and accomplished through application of the logistics process—is guided by a set of overarching principles. Each plan, action, organization, report, procedure, and piece of equipment may be defined and measured in terms of these principles. Each logistics decision is guided by the application of these principles. They are applicable to all military logistics, and provide the common foundation of joint and naval logistics doctrine.

Responsiveness. Providing the right support at the right time and at the right place. This is the most important principle of logistics, because it addresses the effectiveness of the logistics effort; in war an ineffective effort leads to defeat. Ensuring that adequate logistics resources are responsive to operational needs should be the focus of logistics planning. Such planning requires clear guidance from the commander to his planners. It also requires clear communication between operational commanders and those who are responsible for providing logistics support.

Simplicity. Avoiding unnecessary complexity in preparing, planning, and conducting logistics operations. Providing logistics support is not simple, but plans that rely on basic systems and standardized procedures usually have the best chance for success.

Flexibility. Adapting logistics support to changing conditions. The dynamics of military operations are such that change is both inevitable and rapid. Logistics must be flexible enough to support changing missions; evolving concepts of operations.; and shifting tactical, operational, and strategic conditions. A thorough understanding of the commander's intent enables logistics planners to support the fluid requirements of naval operations.

Economy. Effective employment of logistics support assets. Logistics assets are allocated on the basis of availability and the commander's objectives. Effective employment requires the operational commander to decide which resources must be committed and which should be kept in reserve. Additionally, the commander may need to allocate limited resources to support conflicting requirements. The prioritization of requirements in the face of limited forces, materiel, and lift capability is a key factor in determining the logistics feasibility of a plan.

Attainability. The ability to acquire the minimum essential logistics support to begin operations. The difference between this minimum essential level of support and the commander's desired level of support determines the level of risk inherent in the operation from a logistics viewpoint. The accurate determination of the minimum requirements, and the time it will take to reach that level given the available resources, allows the commander to determine the earliest possible date for the commencement of operations.

Sustainability. Ensuring adequate logistics support for the duration of the operation. Sustaining forces in an operation of undetermined duration and uncertain intensity is a tremendous challenge. Forces may operate with a diminished level of support for some time, but every means must be taken to maintain minimum essential support at all times. Sustainability derives from effective planning; accurate projections of requirements; careful application of the principles of economy, responsiveness,

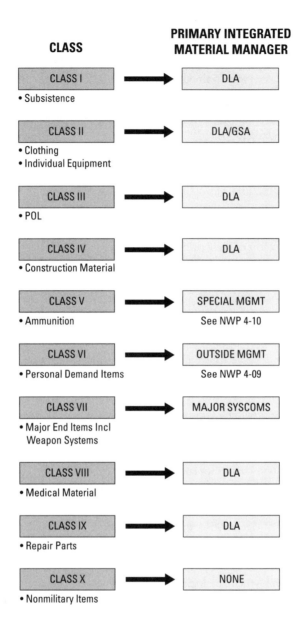

CLASS	PRIMARY INTEGRATED MATERIAL MANAGER
CLASS I • Subsistence	DLA
CLASS II • Clothing • Individual Equipment	DLA/GSA
CLASS III • POL	DLA
CLASS IV • Construction Material	DLA
CLASS V • Ammunition	SPECIAL MGMT See NWP 4-10
CLASS VI • Personal Demand Items	OUTSIDE MGMT See NWP 4-09
CLASS VII • Major End Items Incl Weapon Systems	MAJOR SYSCOMS
CLASS VIII • Medical Material	DLA
CLASS IX • Repair Parts	DLA
CLASS X • Nonmilitary Items	NONE

and flexibility to provide required support; and successful protection and maintenance of the lines of communication.

Survivability. Ensuring the functional effectiveness of the logistics infrastructure in spite of degradation and damage. Logistics forces, sites, transportation modes, lines of communication, and industrial centers are all high-value targets that must be protected. Logistics ships, aircraft, vehicles, and bases may be vulnerable to direct attack by enemy forces or terrorists. Similarly, these assets and the systems that utilize them are subject to disruption by natural disaster, weather, communications failures, civil unrest, contract and labor disputes, legal challenges, and the political decisions of other nations. Survivability requires a robust and diverse logistics system capable of sustaining forces in the face of any obstacle. Dispersion of installations and materiel, maintenance of alternate modes of transportation and lines of communication, redundant logistics communication systems, adequate stock levels, reserves of equipment and personnel, phased delivery, effective use of deception operations, and alternate sources of supply can all support survivability. Force reconstitution and replacement, decontamination, reconstruction, re-equipment, repair, or relocation may restore the effectiveness of logistics systems degraded by battle damage or other events. Accordingly, the survivable logistics must include sufficient assets to support its own recovery as well as the operating forces.

Levels of Logistical Support

Logistics support is provided at the strategic, operational, and tactical levels, and involves interrelated and often overlapping functions and capabilities.

Strategic Logistics encompasses the ability to deploy and sustain forces executing the national military strategy whenever and wherever. It involves determination of requirements, personnel and materiel acquisition, and management of strategic airlift and sealift for the optimum levels of readiness at best value to the Navy. It also includes the role of prepositioned equipment and materiel—both afloat and ashore—and our national ability to maintain the required support levels for the duration of operations. A particular concern at the strategic level is that our industrial bases maintain the capability, capacity, and technology to support timely production of modern weapon systems, support equipment, health services, munitions, stores, and command and control system components to meet wartime requirements. The greater the scope or duration of anticipated military operations, the greater the impact of continuing effective strategic logistics operations.

Operational Logistics involves coordinating and providing theater logistics resources to operating forces. It includes support activities to sustain campaigns and major operations within a theater and is the level at which joint logistics responsibilities and arrangements are coordinated. Operational logistics encompasses theater support sites and activities, ashore or afloat, and the theater transportation required to move personnel and materiel to and from supported forces. It also entails management and protection of those assets. It is the bridge that translates strategic logistics capability into tactical logistics support. The unified combatant commanders and the supporting service component commanders are the main benefactors of this level of logistics.

Tactical Logistics focuses on support within and among combat forces. Navy tactical logistics encompasses the logistics support of forces within a battle group or amphibious readiness group and within Navy elements ashore, from both afloat platforms—including Combat Logistics Force (CLF) ships—and shore-based logistics support facilities. Tactical logistics support activities include maintenance, battle-damage repair, engineering, fueling, arming, moving, sustaining, material transshipment, personnel, and health service. Marine Corps tactical logistics, including combat service support (CSS), is provided by task-organized combat service support elements that complement the organic capabilities of the combat elements.

The Funcitional Areas of Logistics

Logistics activities at each level of support require a broad range of skills, knowledge, and capabilities. These form six major functional areas allowing us to understand, organize, and execute logistics. They are supply, maintenance, transportation, engineering, health services, and other logistics services. Applied in appropriate combination, they provide forces with total logistics support. These functional areas are consistent throughout the Armed Services, and provide a common fabric of logistics organization that facilitates joint operations. Below is a brief synopsis of these respective functional areas of logistics.

Supply provides materiel and services for our forces. The supply function includes design, procurement, contracting, receipt, safe storage, inventory control, issuance, retrograde, and disposal of end items including repairables and consumables. The defense supply system, which includes the Navy and

Marine Corps supply systems, equips and sustains our military forces during all phases of preparation and employment. The defense supply system manages millions of items, which are grouped into 10 classes of supply for management purposes.

Maintenance entails all actions necessary to preserve, repair, and ensure continued operation and effectiveness of systems (e.g., ships and aircraft), components, and equipment. It includes the policy, organization, and activities related to the maintenance of equipment, afloat and ashore. The Marine Corps identifies eight functions of maintenance: inspection and classification; servicing, adjusting, and tuning; repair; modification; rebuilding and overhaul; reclamation; recovery and evacuation. Maintenance strategies, standards of performance for preventative and corrective maintenance, technical engineering support, and battle-damage repair are important components of the maintenance function. Conservation, reutilization, and disposal are also important to the economical and environmentally sound support of forces. While maintenance is primarily involved in the sustainment process, the collection, analysis, and reporting of materiel maintenance data is critical to effective acquisition. Maintenance is conducted at three levels—organizational, intermediate, and depot.

Transportation provides for the movement of units, personnel, equipment, and supplies from the point of origin to the final destination. This function includes deployment and redeployment of supported and supporting forces, the transportation of sustainment resources, movement of forces and resources to ports of embarkation, and inter-theater and intra-theater operations. The transportation system operates at every level of logistics and provides for the movement of casualties, mail, and other critical services as well. The Marine Corps also identifies embarkation, landing support, motor transport, port and terminal operations, air delivery, material handling equipment operations, and freight or passenger transportation as functions of transportation. Further delineation of the levels of transportation include:

Strategic Transportation encompasses the movement of resources to and from the theater of operations. Navy ships and their embarked forces including naval air squadrons and detachments and Marine Corps expeditionary units are initially self-deploying. Sustainment for these forces is provided either by onboard organic resources or via Combat Logistics Force (CLF) ships. Methods to deploy these units are via strategic common-user land, sea, and air transportation provided through the U.S. Transportation Command (USTRANSCOM), utilizing the assets of the Military Sealift Command (MSC), the Air Mobility Command (AMC), and the Military Traffic Management Command (MTMC). These commands use both military and civilian assets as available and appropriate. The Navy's MSC provides the DOD strategic heavy lift and also supports the Marine Corps Maritime Preposition Force (MPF) Program and assault follow-on echelon as well as the U.S. Army's afloat prepositioning needs.

Operational Transportation. This transportation is the bridge between the strategic lift provider and the operating forces. CLF ships, and Navy vertical onboard delivery (VOD) helicopters and carrier onboard delivery (COD) aircraft, provide transportation to and from afloat forces.

Tactical Transportation. Within the battle force, most tactical transportation is via ship's organic aircraft. Assigned helicopters shuttle personnel, sustainment, mail, and other materiel from sites and support ships, and within the battle force. Tactical transportation also allows battle group commanders to share resources and capabilities to enhance the overall readiness of the force. Ashore, sites and forces utilize organic vehicles or aircraft for tactical movements.

Engineering provides construction, damage repairs, combat engineering, and facilities maintenance ashore, executed by Navy, Marine Corps, other Service engineer units and civilian contractors. The Naval Construction Force (NCF) units, known as "Seabees," support Marine Corps engineer capabilities providing extensive technical and manpower resources in constructing advance bases, upgrading supply routes, developing aviation support facilities, and providing battle damage repair.

Health Services support the health of naval personnel and their families. This support includes medical and dental materiel, facilities, and services in both combat and non-combat situations. In contingency operations, these services are provided through organic assets including hospital corpsmen, shipboard sick bays, medical and dental battalions of the Marine Corps' FSSG, Fleet Hospitals, hospital ships (T-AH), and other fixed outpatient and inpatient facilities including other-Service, contract, or host nation facilities. The functions of HSS are health maintenance, casualty collection, casualty treatment, temporary casualty holding and evacuation, emergency and routine health care, monitoring the health, sanitation, and medical readiness of deploying forces, medical service record maintenance, and maintaining mass casualty plans.

Other Logistics Services are required to provide administrative and personnel support to achieve maximum operational capability of a force. This support extends to those areas of personnel support, quality of life, and morale issues that help define the combat effectiveness of the individual. Other logistics services include billeting, disbursing, exchange services, food services, legal services, morale, welfare, and recreation (MWR), mortuary affairs, and postal services.

These six logistics functional areas combine and integrate to provide total logistics support. Planning and execution of responsive, sustainable support requires balancing the functional areas to provide the right support at the right time and place. The appropriate balance and level of support flow through various activities, channels, modes, and nodes to the end user. Regardless of the type of support or the specific means of delivery, logistics support across the full range of functional areas is provided through a series of elements.

Logistical Planning

Naval logistics planning and information support is designed to answer these questions: What materials, facilities, and services are needed? Who is responsible for providing them? How, when, and where will they be provided? To find answers, we start with sources of logistics planning guidance, then apply a formal process that parallels operational planning procedures. The nature of the situation will determine whether we apply a deliberate or a crisis action planning process. These rules and tools are imperative for optimizing logistics systems responsiveness to the warfighter. Naval logistics information support systems keep the plan current, accurate, and adequate by providing data on the status of logistic resources, operational force needs, and the ability to meet those needs. Logistics planning and information support are thus complementary.

THE DELIBERATE PLANNING PROCESS

Deliberate planning prepares for a possible contingency based on the best available information, using forces and resources apportioned by the CJCS in JSCP. Most deliberate planning is done in peacetime, based on assumptions regarding the political and military circumstances that may prevail when the plan is implemented. Deliberate planning is highly structured and occurs in regular cycles. It produces an operations plan (OPLAN), concept plan (CONPLAN), or functional plan.

Logistics planners prepare the staff logistics estimate during the concept development phase of deliberate planning. This provides the commander with the information to support courses of action (COA) selection, and is developed concurrently with the commander's estimate. Logistics also plays a major role in plan development as supported and supporting CINCs determine support requirements and resolve shortfalls. Finally, logistics planners at many levels prepare supporting plans to provide the mobilization, deployment, sustainment, reconstitution, and redeployment of forces and resources in the OPLAN.

CRISIS ACTION PLANNING

Crisis Action Planning (CAP) is conducted in rapid response to actual circumstances. CAP follows the general pattern of deliberate planning, but adds flexibility for timely action. If an existing OPLAN is adaptable to the situation, CAP procedures are used to adapt an existing OPLAN to actual conditions or to develop and execute an operation order (OPORD).

Considerations. Logistic planners identify and resolve support problems early by working concurrently with, and in support of, operations planners. Detailed logistics planning should:

- Achieve optimum warfighter readiness.
- Optimize logistics systems responsiveness to the warfighter.
- Earmark significant time-phased support requirements necessary to maintain and sustain the warfighter whenever, wherever.
- Identify personnel and cargo throughput at shore-based logistics sites.
- Identify transportation requirements to support the movement of personnel, equipment, and supplies.
- Outline the capabilities and limitations of ports, including the Logistics-Over-the-Shore (LOTS) capability to respond to normal and expanded requirements.
- Recognize support methods and procedures required to meet the needs of the sea, air, and land lines of communications.
- Coordinate and control movement into the contingency area.
- Develop reasonable logistical assumptions.
- Define the extent of needed host nation resources.
- Designate alternative support sources for host nation support failure.
- Identify the engineering and construction requirements for sustainability.
- Identify the source of funding for logistics support.
- Delineate contracting responsibilities and authority.
- Consider the meteorologic and oceanographic limitations.
- Identify health service support requirements.
- Identify the service and maintenance support requirements for sustainability.

REVIEW QUESTIONS

1. Describe the seven principals of logistics.
2. Explain how the levels of war and logistical support interact.
3. Identify the different areas required for effective logistical support.
4. Explain the supply planning processes.

CHAPTER 8

Surface Warfare

LEARNING OBJECTIVES

At the end of this lecture the student will be able to:

- Describe the principal objectives of SUW.
- State the command relationship between the CWC and the SUWC.
- Describe organization of GCCS-M.
- Describe the concept of a Surface Strike Group (SAG).
- Describe the various platforms and weapons involved in SUW.
- Describe the four phases of SUW.
- Describe the considerations when using Over-the-Horizon Targeting.
- Describe the problems associated with Over-the-Horizon Targeting.

Introduction

The days of close engagements between battleships are long gone and it has been proven that air supremacy is not enough to render surface ships ineffective. The improved air defense capabilities of ships coupled with cruise missile technology requires a coordinated effort between TACAIR, surface ships, and submarines to conduct effective Surface Warfare (SUW) engagements. SUW, formerly called Anti-Surface Warfare (ASUW), is the destruction or neutralization of enemy surface combatants and merchant ships.

Nonetheless naval forces are inherently flexible, lending themselves to varying employments and degrees of application of force. Thus, a wide variety of tactical employment is possible, ranging from humanitarian assistance through strike operations to extended joint operations ashore. Shifting the strategic focus to regional challenges moves many mission types from the periphery to the forefront of concern. Though each mission will be different, there are common considerations for the tactical commander in planning. This chapter covers these considerations as they relate to the tactical employment of surface ships.

TACTICAL MISSIONS

Tactical missions tasked to surface combatants fall into six categories:

1. Littoral strike
2. Sea control
3. Escort
4. Police
5. Special
6. Coordination

Littoral Strike Missions

Strike warfare is the attack of land-based targets by sea-based forces. This power projection role is increasingly important in the littoral areas of the world.

Power projection ashore can be accomplished by:

1. Ballistic missiles from submarines
2. Attack aircraft from aircraft carriers
3. Cruise missiles from surface combatants and submarines
4. Shore bombardment from surface combatants
5. Marine amphibious assault, or follow-on joint expeditionary warfare, from AMW ships and strategic sea and airlift
6. Special forces, deployed by surface ships, submarines, or aircraft

Surface combatants will likely be involved in all of these littoral strike missions except ballistic missile strike.

CRUISE MISSILE STRIKE

Most cruisers, SSNs, and destroyers have strike-capable cruise missile systems. Cruise missile strikes from surface combatants can be conducted independently or integrated with other types of strike. For example, a strike of cruise missiles from combatants and tactical attack aircraft from a carrier is an effective combination. Such a strike could be conducted entirely from a CSG that has cruise missile ships, or as a result of coordination between a CSG and a strike SAG, as part of a battle force. A strike SAG might consist of two to six ships, divided into two or three TUs. Maximum stealth in approach is used, utilizing the island characteristics of the littoral, dispersed forces, maximum control of emissions, high speed, and diversionary routes.

GUNFIRE IN SHORE BOMBARDMENT

Shore bombardment is the attack of land targets using gunfire that is not associated with an amphibious operation (the latter is called naval gunfire support (NGFS)). Cruisers and destroyers can conduct shore bombardment with 5-inch guns at ranges of over 10 nm. The primary tactical grouping for shore bombardment is the strike SAG. Shore bombardment by surface combatants can be conducted independently or integrated with other types of strike. For example, shore bombardment of a target on an island followed by cruise missile launch at a continental target could be conducted, both integrated with strikes by tacatical attack aircraft. If the target is visible or can be detected using line-of-sight radar, the ship can fire at it without offboard assistance. This is known as direct fire.

If the target is not visible, indirect fire is employed. Shore bombardment does not normally employ offboard spotting. The ship fires where intelligence indicates there is a target. Offboard target identification and spotting may be used if available and the threat permits. Such support may come from:

1. Special operations units ashore
2. Nonorganic aircraft
3. Organic aircraft (LAMPS helicopter)
4. Allied or friendly guerrilla forces
5. UAVs

Shore bombardment is primarily useful against the following types of targets:

1. Direct fire:

(a) Coastal roads
(b) Coastal logistic convoys
(c) Coastal artillery sites
(d) Coastal ports and bases
(e) Ships and submarines in port

2. Indirect fire:

(a) Large area logistic depots
(b) Airfields
(c) Aircraft on the ground
(d) Fuel depots
(e) Ammunition depots
(f) Troop concentrations

Missions for shore bombardment depend on whether ships can close to the necessary range. Typical missions for shore bombardment are:

1. Destroying ships, submarines, and patrol craft in their bases
2. Interdicting logistic movement along coastal routes
3. Neutralizing shore defenses
4. Destroying logistic infrastructure supporting enemy forces
5. Destroying political targets that influence the war making will of opposing leaders
6. Neutralizing ports in support of sea control
7. Diversionary attack

An SAG conducting shore bombardment would employ approach and retirement tactics similar to those for TLAM attack but at closer range.

NAVAL SURFACE FIRE SUPPORT (NSFS)

NSFS is conducted in association with expeditionary warfare and comprises all methods of attacking shore targets with surface ship weapons systems. These methods include naval gunfire, missiles, rockets, and EA delivered by naval surface forces in support of amphibious operations. NSFS is normally conducted by surface combatants assigned as escorts of an amphibious task group (ATG). Strike SAGs and sea control SAGs also have NSFS capability. Such groups could be assigned to an ATF for NSFS as well as for defense of the AOA.

NGFS

NGFS is conducted in association with expeditionary warfare and is a form of NSFS. Support of ground forces using naval gunfire is provided in the prelanding, landing, and postlanding phases of an amphibious operation. NGFS can also support joint U.S. or combined U.S. and allied forces operating in littoral areas.

NGFS is normally conducted in one of the following modes:

1. Direct fire
2. Call for fire
3. Harassment and interdiction fire

Direct fire is targeted and spotted from the ship at targets of opportunity or emergent threats. Call for fire is provided at the request of a spotter ashore. The spotter provides target coordinates and adjusts the fall of shot through communications with the firing ship. Call for fire can be carried out against a point target (e.g., a bunker) or an area target (e.g., massed troops).

Harassment fire subjects enemy troops to random fire over a period of time. Interdiction fire saturates an area with random fire over a period of time to deter transit or occupation. Indiscriminate use of harassment and interdiction fire may cause excessive collateral damage. NGFS is conducted in a manner similar to shore bombardment; however, it is normally conducted within directed fire support areas (FSAs) where sea control has been established and the threat to fire support ships is low. NGFS ships are normally assigned stations to support a corresponding area ashore and remain in that area unless a threat arises or they are reassigned.

EXPEDITIONARY WARFARE SUPPORT

Surface combatants may employ cruise missile strikes, shore bombardment, NSFS, or special warfare strikes to support amphibious operations or follow-on joint operations along the littoral. They may be assigned to escort an ARG or strike ARG and, when landings commence, may be assigned to defend the AOA (air defense zone coordinator (ADZC) operations).

SPECIAL WARFARE MISSIONS SUPPORT

Special forces may be used to conduct missions against selected targets ashore. Such missions may be independent or in conjunction with the approach of a battle force, strike force, or landing force. Surface combatants may be assigned to support special warfare strikes with:

1. Covert approaches to insert or withdraw special forces
2. Sea control operations in support of landings or withdrawals
3. Diversionary (OPDEC) maneuvers in support of landings or withdrawals

Sea Control Missions

Sea control missions are often a prerequisite to the projection of power. Sea control missions can include:

1. Operations in support of offensive strikes
2. Offensive attrition of enemy naval forces in the open seas
3. Offensive and defensive operations to control chokepoints
4. Offensive and defensive minelaying operations
5. Mine countermeasures (MCM) operations
6. Attack on enemy SLOCs
7. Defense of the U.S. maritime defense zone (MDZ)
8. Defense of U.S. and Allied SLOCs
9. Defense of overseas bases

SEA CONTROL IN SUPPORT OF OFFENSIVE STRIKES

To minimize losses in a tactical strike, it is necessary to establish sea control where the strikes will be launched from. Such sea control involves:

1. Air superiority
2. Surface superiority
3. Subsurface superiority
4. Control of space assets affecting the area
5. Control of the electromagnetic spectrum in the area

Sea control in support of offensive strikes may be established for a brief period or an extended period of time, depending on the nature of the mission. Examples of different types of sea control are:

1. Sea control in support of tactical aircraft strike. Sea control is necessary in the objective area for a period corresponding to the time required to:

   (a) Transit to within aircraft range of the target
   (b) Launch aircraft, deliver the strike, and recover aircraft (for the number of strikes planned)
   (c) Withdraw from the primary threat area

2. Sea control in support of TLAM strikes. This is the same as for a tactical aircraft strike. If strike ships must be ready to launch on short notice, they may need to remain in a relatively restricted TLAM launch basket for an extended period.

3. Sea control in support of shore bombardment. Same as above, but requires operations close to enemy shores and a high degree of sea control for a very limited period. Normally conducted as an "in- and-out" operation.

Sea control in support of offensive strike operations can be conducted using the following methods:

1. Rollback—Forces operate at the edge of enemy-controlled seas and operate to:

(a) Lure enemy forces into areas where own forces are equal or superior and enemy forces can be destroyed in detail

(b) Gradually destroy enemy forces and push back their area of sea control to permit strike operations.

2. Raid—Maneuver and OPDEC are used to penetrate enemy-controlled seas with minimum need to fight on the way in. Enemy forces are then engaged, some attrition is achieved, and local sea control is established for a limited time before withdrawal.

OFFENSIVE ATTRITION OF ENEMY NAVAL FORCES IN THE OPEN SEAS

Enemy forces in the open seas may include surface, submarine, and land- or sea-based air forces. Their objectives are likely to be:

1. Challenge U.S./allied sea control
2. Establish enemy sea control
3. Attack SLOCs
4. Establish local sea control in support of strike operations

In most conflicts, general or limited, destruction of enemy naval forces is a primary objective. Therefore, naval forces, including surface combatants, may be assigned a primary mission of attacking enemy forces in the open seas. This type of mission may include:

1. Barrier operations —Establish an open sea barrier through which enemy forces must pass to achieve position in the open sea. The barrier is used to detect and then conduct offensive attack against transiting enemy forces. For example, a sea control SAG with organic helicopters could establish a barrier against surface forces or submarines.
2. Area attack operations—A sea control SAG could maintain a designated area of open sea free of enemy forces.
3. Specific attack operations —A sea control SAG could be assigned to counter a specifically designated enemy force known to be operating in the open sea.

OFFENSIVE AND DEFENSIVE OPERATIONS IN CHOKEPOINT CONTROL

Chokepoints are important tactical objectives because the force occupying them can control the passage of naval forces, sealift, and merchant shipping. Chokepoints are normally within or partially within the territorial sea of various nations, so alliances, treaties, and political considerations enter into their closure or control. Examples of chokepoints are:

1. Straits of Gibraltar
2. Panama Canal
3. Straits of Malacca
4. Straits of Hormuz
5. Florida Straits

The commander should carefully analyze the chokepoints that could contain an enemy and plan to control them. Types of control operations are:

1. Offensive chokepoint closure—An SAG may be assigned to close a chokepoint to prevent enemy forces from reaching tactical position and to cause attrition.

2. Defensive chokepoint closure—An SAG may be assigned to prevent enemy forces from approaching an area to be defended.

OFFENSIVE AND DEFENSIVE MINELAYING OPERATIONS

Mines are powerful yet inexpensive. A minefield is an effective way to cause attrition of enemy forces; it is also an effective deterrent against enemy forces using an area that has been mined. Surface combatants and civilian vessels can lay by aircraft, submarines, logistics ships, and with equipment modifications, mines. U.S. surface combatants are more usually employed to protect forces engaged in minelaying. Types of minelaying missions are:

1. Offensive—Mines are laid near enemy ports, bases, and in chokepoints to control access to the sea. Such minefields, located where the enemy may exercise sea control, are normally laid by submarines and aircraft.
2. Defensive—Mines are laid along approaches to friendly ports and naval bases to be defended. Information on access routes is promulgated only to friendly forces. Such minefields are designed to prevent the approach of enemy submarines and surface ships within striking range. Another type of defensive field might be one laid in a chokepoint to prevent enemy submarines from controlling it, while allowing friendly forces to transit via known routes.

PROACTIVE AND ENABLING MCM.

The enemy can also use mines against us. Therefore, naval forces require MCM capability. MCM is the aggregation of procedures used by naval forces to prevent loss of ships and personnel to minefields. Proactive MCM involves preventing enemy mining. Enabling MCM is conducted to sweep or reduce the effectiveness of existing minefields. Enabling MCM is conducted by specially equipped MCM ships, surface craft, helicopters, explosive ordnance disposal (EOD) detachments, and naval special warfare units. Typical proactive MCM missions are strike operations against:

1. Mine stockpiles
2. Mine distribution networks
3. Surface, subsurface, and air mine delivery vehicles

Typical types of enabling MCM missions are:

1. Breakout and clearing of harbors where naval and merchant ships transit to keep them clear of mines
2. Clearing the transport areas and the approach and transit lanes for an amphibious assault
3. Exploration/reconnaissance to determine the presence of mines and the characteristics of a minefield
4. Clearing chokepoints

ATTACK ENEMY SLOCS

Most littoral nations depend on seaborne supplies of food, petroleum, natural resources, materials, weapons, parts, and other requirements for modern life and warfare. Therefore, interdiction of critical supplies can reduce an enemy's capability and will to initiate or continue a conflict.

Supplies are carried in merchant ships that are usually limited in number and availability. Attrition of merchant ships at a rate faster than they can be re-placed will reduce an enemy's war-making capacity. Sea control SAGs and submarines are best suited for this mission. While the larger, more capable CSGs and strike SAGs are engaging enemy combatants, smaller commerce raider SAGs can attack enemy SLOCs. This can be accomplished while engaged in other operations.

DEFENSE OF THE MDZ

U.S. naval strategy is to establish an offensive, forward-deployed defense that acts as a deterrent in low levels of conflict and starts from an advantageous tactical position if a conflict escalates. This strategy holds conflicts at arm's length from the U.S. homeland. Nevertheless, some enemy forces, primarily submarines, may leak through into U.S. coastal waters. Such submarines would likely have missions such as:

1. Attrition of merchant shipping
2. Attrition of naval combatants near bases
3. Mining of harbors and chokepoints
4. Diversion of U.S. assets from offensive operations
5. Shock, psychological, and propaganda effects
6. Missile launch against military targets ashore
7. Missile launch against industrial, political, and population centers

Surface combatants and MCM forces will normally be assigned to operate defensively in the MDZ in coastal waters. SAGs could operate in the MDZ with missions such as:

1. Offensive ASW
2. Escort of coastal ships and convoys
3. Search of high probability areas for enemy submarines
4. Defense of MCM forces

DEFENSE OF U.S. AND ALLIED SLOCS

The U.S. depends on seaborne transit for strategically important natural resources. Our economy is built around the ability to trade freely. U.S. alliances are built around maintenance of SLOCs with our trading partners. A primary sea control mission is therefore defense of U.S.–allied SLOCs. This mission is best performed by a combination of forces, of which surface forces are a vital component.

Escort Missions

In escort missions, surface combatants protect a large ship or group of ships whose capability is essential to the tactical objective. Missions may include escort of:

1. An aircraft carrier in a CSG or expeditionary strike force (ESF)
2. A mix of ships in a sea control ARG
3. Amphibious ships in an ATG and joint expeditionary follow-on forces
4. Maritime prepositioning ships (MPS)
5. Logistics ships in a combat logistics force
6. Strategic sealift in a convoy
7. Special assignments

Surface combatants are multicapable and can be assigned secondary missions within an escort mission. For example, a destroyer escorting a CSG may also have a secondary mission of cruise missile strike integrated with a strike by the attack wing.

AIRCRAFT CARRIER ESCORT IN A CSG OR ESF

The aircraft carrier is a floating airfield and its aircraft greatly enhance the capabilities of surface forces. Surface combatants are assigned as escorts for a CSG. Cruisers and destroyers are normally assigned this mission; however, frigates may also be assigned as available. CSG escorts are normally assigned to the following duties:

1. Inner screen ships defend against close-in submarine torpedo attacks and surface raiders, and provide close-in missile defense against aircraft and antiship missiles.
2. Outer screen ships provide: long-range detection and attack of submarines, particularly those capable of firing antiship missiles; air search alert for the main body and for own ship's defense; surface search alert for the main body; and defend against surface raiders.
3. Picket ships provide very long-range detection and attack of enemy aircraft along a threat axis and local defense against submarines and surface raiders.
4. Towed array ships placed in advanced towed array patrol areas provide ASW detection along a threat axis.
5. Logistics ship escorts defend combat logistics force ships assigned to the CVBG.
6. SAUs detached from the CVBG under a designated SAU commander to conduct ASW operations. AAW defense of the SAU is provided by missile ships in the SAU or fighter aircraft assigned to the SAU commander for control.
7. ASUW SAGs detached to conduct advance ASUW operations, including marking or attacking enemy surface units.
8. Strike SAGs detached to conduct littoral strike missions including cruise missile strike, shore bombardment, or NSFS.
9. ADZC consisting of an AEGIS or new threat upgrade (NTU) missile ship detached to protect the AOA or area of joint expeditionary operations.

Normally, four to nine surface combatants are assigned to one carrier as a CSG. It is tactically difficult to use this number of ships in screen formations that adequately cover wide sectors. Therefore, orient ships in layers along a primary threat axis, based on the intelligence available. Maximum use of maneuverability, speed, and OPDEC is desirable.

When possible, escorts should employ offensive tactics along the threat axis, while main body units transit using high speed, maneuver, and OPDEC. This is particularly so when transiting through open ocean or sea. When a CSG approaches a chokepoint, three tactical actions are necessary:

1. Use advance TUs with air support to clear the way.
2. Concentrate the screen tighter than in the open ocean or sea.
3. Use OPDEC and tactical geography to shield approach and transit.

SEA CONTROL ARG ESCORT

The sea control ARG is a mission-defined, asset-driven tactical grouping formed to operate as a sea control tactical group in the absence of a CV. It usually consists of an LHA or LHD with AV-8 wing, associated amphibious ships, four to six surface combatants (DD/DDG, CG, FFG), and possibly SSNs. Tactics to be employed in escort of a sea control ARG are similar to those for a CSG. Because the LHA/LHD air wing is much more limited, a sea control ARG is more reliant on helicopters, space- and land-based sensors, and land-based aircraft. Escorts should include very long-range missile ships to offset the lack of fighter aircraft.

AMPHIBIOUS SHIP ESCORT IN AN ATF AND JOINT EXPEDITIONARY FOLLOW-ON FORCES

Escorts assigned to ATFs will usually participate in all phases of an amphibious operation:

1. Planning
2. Embarkation and port breakout
3. Assault rehearsal(s)
4. Transit
5. Securing the AOA prior to arrival of amphibious ships
6. AOA defense during ship-to-shore movement
7. AOA defense and NSFS during beachhead consolidation (may be prolonged into a period of sustained joint operations)
8. Withdrawal of amphibious shipping. All phases of an amphibious operation may be supported by a CSG strike SAG, sea control ARG or sea control SAG as part of a larger battle force. During embarkation and port breakout, surface escorts will normally be employed in primary ASW missions. Transit of cleared routes through minefields may be required.

In the transit phase, surface combatants employ the tactics described for CSG transit. The primary difference is that there are more ships of the main body to protect. Therefore, screening may require more emphasis than advance offensive tactics. Advance offensive tasks may have to be accomplished by other supporting forces.

The transition between transit and assault is a critical phase. Surface, air, subsurface, and mine threats must be reduced before amphibious ships arrive in the AOA. Small, high-speed patrol craft mixed with fishing boats present a particularly difficult challenge for screening ships.

AOA defense during both the ship-to-shore movement and beachhead consolidation phases, usually involves screening a geographic area around the objective area. Like all screens, this will be more effective if supplemented with long-range picket ships and advance TUs operating offensively around the screen perimeter. When amphibious tactics include over-the-horizon launch, escorts for the launching ships should operate independently from those in the fixed screen in the AOA. Depending on the air and high-speed surface threats, the route to be taken by amphibious craft (such as air cushion landing craft (LCAC)) may also require protection.

Normally, an ATF includes ships with embarked transport and attack helicopters, and when an LHA or LHD is present, vertical/short takeoff and landing (V/STOL) fighter/attack aircraft. At CATF's and CLF's discretion, these aircraft may be used to enhance the capabilities of surface combatants during all phases of the operation. Recognize, however, that the aircrafts' primary purpose is to support the amphibious operation and that there will be a need to preserve them for that purpose.

Following the establishment of a secure beachhead ashore, joint or combined forces may pass through the AOA to build up ground and air combat power for a period of sustained operations. This is essentially an extension of the beach consolidation phase which gradually passes into heavy sealift support. Escort tactics derive from and resemble defense of the AOA.

MPS ESCORT

MPS are logistics ships preloaded with equipment and supplies to support Marine Corps forces. In a crisis, Marine forces are flown to an airfield near an objective area while the MPS proceed to the nearest port. The Marines then join up with their equipment and proceed to carry out their mission. MPS provide rapidly deployable, medium to heavy forces for regional crises. They can be prepositioned anywhere friendly ports or anchorages are available. MPS can be used:

1. In a benign environment where port entry or the use of an offload beach is not challenged
2. In a hostile environment where port entry or the use of an offload beach is not welcomed, but sufficient force to challenge it does not exist
3. To follow an opposed amphibious landing with additional equipment and supplies. Surface combatants are not normally assigned as MPS escorts unless the threat is high enough to warrant it. This threat might be:

   a. An at-sea threat to MPS transit (from patrol boats, ships, aircraft, submarines, or mines)
   b. A land-based threat to MPS operations. Tactics for MPS escort are similar to those for escort of an ATG. The primary difference is that, while amphibious ships have embarked aircraft, small caliber guns, and AAW point defense systems, MPS have no self-defense capability.

LOGISTICS SHIP ESCORT

Logistics ships are used tactically in two ways:

1. Logistics ships accompanying battle groups
2. Shuttles between ports and the battle force to keep the force's logistics ships replenished. The battle force commander may assign surface combatants to escort logistics ships employed in either way. Tactics are similar to those used in the transit phase of an amphibious operation.

STRATEGIC SEALIFT ESCORT

Protection of sealift is a traditional naval mission that retains its importance in a littoral strategy. Our nation must remain able to deliver heavy equipment and resupply major ground and air combat power forward in crisis. The scope and nature of the crisis will determine the degree of protection required by sealift.

In general war, or war against a powerful regional adversary, sealift (merchant ship) transits would be controlled under the Navy control of shipping (NCS) concept. Convoys are normally employed to provide efficient defense. In a regional crisis, ocean transit might be individual and unescorted, but the approach phase might be convoyed and escorted for protection from land-based air forces, shore-based missiles, diesel submarines operating in shallow water or at chokepoints, and mines. Surface combatants are employed as CEGs. The size of a group depends on the threat and the size of the convoy, but would normally be five to eight ships.

Frigates are designed for convoy escort, but cruisers and destroyers may also be used. Convoy escorts must be prepared for port breakout and break-in operations and for countering air, submarine, surface raider, and mine threats. Tactics are similar to those used in escorting an MPS group. The primary difference is that the escort-to-escorted ratio is higher in a convoy than in other tactical groupings.

SPECIAL ESCORT ASSIGNMENTS

Since surface combatants are general purpose fighting ships, they can expect special escort assignments, including escort of:

1. Special forces for covert operations (they may transport them)
2. Politically sensitive ships (hospital ships, special cargos)
3. Valuable ships under tow
4. Ships diverted from maritime interception zones
5. Captured ships

Police Missions

Police missions are associated with law enforcement on the high seas and within U.S. coastal waters. Police missions are frequently carried out with, and in support of, the Coast Guard. Examples of police missions include:

1. Antiterrorist or piracy operations suppression
2. Offshore resources protection
3. Antismuggling and counterdrug (CD) operations
4. Show of force

ANTITERRORIST OR PIRACY OPERATIONS SUPPRESSION

This is the oldest mission of surface ships and the reason regular naval forces were first constituted. Piracy is increasing in many unsettled regions of the world and is still a definite part of the police mission of naval forces. Surface combatants are well suited to counter piratical or terrorist operations at sea or to impose blockades or sanctions on ports or states where such acts are ongoing or have been tolerated.

Tactics to employ in countering piracy include:

1. Signaling a ship to halt flag or light signals, voice radio, loudhailer, or (if authorized) warning shots.
2. Boarding and searching (using boats and helicopters).
3. Tactical gunfire to disable a ship with minimum damage or casualties.
4. Forced boarding. Normally, a surface ship assigned to a police mission is augmented with specially trained personnel who carry out the boarding, but it is possible in limited situations to board with ship's company. Ships can be boarded in the following ways:

   (a) Alongside boarding
   (b) Boarding by armed helicopter, supported by ship guns and small arms
   (c) Boarding by armed boat, supported by ship guns and small arms and armed helicopter, if available
   (d) Boarding by swimmers, overt or covert, supported by helicopters and boats

5. Coercing an unboarded ship to proceed as directed to a designated location or port.
6. Communications jamming to prevent enemy coordination with shore-based support. Single ships or small SAGs are adequate for most antipiracy operations. Tactics employed to interdict ports in which terrorists or pirates are active include:

a. Blockade or quarantine of specific items or supplies
b. Containment of terrorist or pirate ships and boats in port
c. Special warfare strikes against selected targets
d. Cruise missile or gun strikes against selected targets

An SAG is generally the correct size unit to operate against a pirate or terrorist port. Against a sophisticated air threat, employ an SAG close in, supported by tactical aircraft from a CSG.

OFFSHORE RESOURCES PROTECTION.

Most nations, including the United States, claim resource, fishing, and mineral rights out to 200 nm from their shores. Within these waters, the following assets may require protection in wartime:

1. Petroleum drilling towers and production platforms
2. Refining facilities and submerged pipelines
3. Fishing boats and fisheries resources (fish)
4. Mining ships and mineral resources

The 200-nm claim may itself require defense against encroachment. Nations lacking treaty rights to the resources in claimed waters may attempt to use them. If they are permitted to operate without challenge, under international law they will gradually gain the right to use the resources. SAGs or individual surface combatants are well suited to this mission. They usually operate in cooperation with the Coast Guard in U.S. waters or allied navies or coast guards in their respective waters.

ANTISMUGGLING AND CD OPERATIONS.

Narcotics trafficking is a highly profitable activity that results in great damage to the U.S. population. Though military forces, including surface naval forces, have no constitutional power of arrest, surface forces can provide significant support to the DOD "Lead Agency." Surface force CD mission is to detect and monitor the flow of illegal drugs into the continental United States Working with law enforcement agencies, surface forces can provide early detection and monitoring of air and maritime drug smugglers moving primarily from transshipment countries in South America to the United States Tactics against Caribbean/Eastern Pacific air targets of interest normally include early detection, transit monitoring, and air control of intercept or aircraft to positively identify the target aircraft, monitor drop zone activity, and/or hand off to law enforcement agencies. Tactics against surface targets of interest normally include control of maritime patrol aircraft (MPA), surveillance, detection, and querying of suspect vessels, and, if warranted, interception to allow embarked U.S. Coast Guard law enforcement detachments to board, search, and apprehend criminals as appropriate. In support of CD detection and monitoring, surface force contributions include:

1. Intelligence collection for early warning of smuggling activities
2. Rapid and accurate detection, sorting, and monitoring and effective handoff to law enforcement forces
3. Providing C2 to effect these missions. Navy surface ships assigned to CD operations usually form task groups reporting to a CD CJTF

SHOW OF FORCE

A show of force at the right place and time may be sufficient to deter a hostile intent. A show of force can be accomplished under several conditions, as follows:

1. Routine presence operations—maintenance of deployed units in politically sensitive areas.
2. Crisis operations—movement of units to an area of specific political crisis to signal the intent to intervene if necessary.
3. Confrontation operations—use of naval force in a crisis when it is confronted by an opposing naval force.
4. Freedom of navigation operations—use of naval forces to demonstrate and enforce free use of the international sea and certain straits and passages.

A show of force is only credible when it is clear that it can carry out operations at any credible level of escalatory conflict. It must also be able to neutralize any expected enemy opposition. Therefore, the size and composition of the tactical group necessary for a show of force depends on the scenario and the locale.

Special Missions

As general purpose, multiwarfare-capable ships, surface combatants can be assigned a variety of special missions, including:

1. Search and rescue (SAR)
2. Assistance at sea
3. Intelligence collection
4. Tattletale and marker operations
5. Countermarker operations
6. OPDEC
7. Maritime interception operations (MIO)
8. Humanitarian assistance
9. NEOs
10. SPMAGTF

SAR

SAR missions are equally important in peace and war. In peacetime, surface combatants may be involved in search operations for lost boats, downed aircraft, and personnel or gear lost overboard. In combat, they may search for survivors where ships are sunk or aircraft lost. In some operations, certain ships may be designated as primary rescue ships.

ASSISTANCE AT SEA

Ships may be assigned to assist another ship or boat that is in danger, either in peace or war. Typical missions include:

1. Firefighting and damage control assistance
2. Towing
3. Medical assistance
4. Repair assistance
5. Logistic assistance
6. Navigational assistance
7. Removal of injured personnel
8. Protective escort of boatlift
9. Turnback operations

INTELLIGENCE COLLECTION

Intelligence collection can be assigned to surface ships at any time. Typical missions are:

1. Gate guard—Maintaining visual, electromagnetic and acoustic watch on ports and chokepoints to detect and report movements.
2. Surveillance—Maintaining visual electromagnetic, and acoustic watch on a designated area, ship, or group, reporting movements, exercises, observed tactics and any unusual events.
3. EW intelligence collection—Specific collection of enemy or potential enemy electromagnetic signals for analysis and exploitation
4. Acoustic intelligence collection—Specific collection of enemy or potential enemy acoustic signals for analysis and exploitation.

TATTLETALE AND MARKER OPERATIONS

A tattletale is a ship whose mission is to remain within visual and radar range of an enemy or potential enemy unit and provide movement and targeting information to other forces. A marker is a tattletale with immediate offensive or obstructive capability of its own. Primary tactics of both tattletales and markers are to remain covert or undetected for as long as possible while shadowing the marked force. Upon commencement of hostilities, a marker must be prepared to conduct an immediate strike, withdraw at high speed, and defend itself.

COUNTERMARKER OPERATIONS

Antitattletale and countermarker operations are extremely sensitive to changes in ROE. Surface combatants are often assigned as countermarkers. Recommended tactics include:

1. Maintain position between the main body and the marker.
2. Prevent peacetime or brink-of-war harassment of other ships by the marker.
3. Prevent the marker from achieving an attack position on the main body.
4. Caution and if necessary prevent the marker from interfering with safe operations if the marker's aim is primarily political or public relations in nature (e.g., a ship belonging to environmental or other activist).
5. Maneuver to deceive the marker or tattletale and cause it to lose track of the main body.
6. Alert the main body of any change of readiness in the marker or communications by the marker or tattletale that indicate preparation for hostility.
7. When permitted by ROE, engage and destroy the marker or tattletale. If possible, knock out weapons and communications equipment first.

OPDEC

Surface combatants may be assigned to carry out OPDEC, with the following objectives:

1. Provide false information to the enemy
2. Mask location of own forces
3. Make the enemy believe own forces are where they are not
4. Make the enemy believe own forces have more or less capability than they have
5. Cause the enemy to concentrate forces in areas to his tactical disadvantage
6. Cause the enemy to commit forces in ways that cause their early detection
7. Cause the enemy to expend ordnance on false targets or targets of small value

Typical surface ship OPDEC missions include operating:

1. As an OPDEC unit, maneuvering separately from the main body to attract enemy attention
2. As a missile trap, luring enemy aircraft away from the main body and within range of ASW systems
3. As a torpedo trap, luring enemy submarines away from the main body and within range of ASW systems
4. As a communications lure, emitting false communications to confuse the enemy
5. As a jamming unit to jam enemy communications and radars
6. Special decoy signal equipment (vans with dedi-cated personnel)
7. Various floating offboard decoy systems

MIO

Interception operations can be carried out under many names, including blockade, quarantine, cordon sanitaire, and turnback operations. While each has distinct legal implications, the common purpose is to limit or stop export or import of certain strategically important goods, items, weapons, or personnel.

Under international law, a blockade is an act of war which clearly demonstrates clear hostile intent. Blockades may be established to keep material out of an area or to confine material or ships to an area. For example, a blockade may prevent enemy ships from exiting a territorial sea and threatening joint operations ashore. Blockades can be close (ships actually in blocking positions) or distant (ships within striking distance). Distant blockade is tactically preferable as it uses fewer resources and results in less wear on ships. Offensive mining now replaces most close blockades. The test of a blockade is effectiveness; a blockade is not legally recognized unless it can be enforced.

Quarantine is a lesser form of control, intended to display intent to control a certain situation, but without hostile intent.

A cordon sanitaire is a political declaration that places a designated area out of bounds for opposing forces. If effective, it deters certain kinds of enemy action in that area and permits friendly operations therein to proceed.

Typical ways to enforce a quarantine or cordon sanitaire are:

1. Threat of escalation
2. Minefields
3. Submarine patrols
4. Surface combatant patrols
5. CSG or sea control ARG
6. Land-based tactical aircraft (area permitting)

At the low end of the spectrum, antismuggling and CD operations, antiterrorist operations, turnback of illegal boatlifts, and other forms of police operations overlap MIO.

MIO typically involve exercising local sea control near designated coastlines, chokepoints, and ports. All forms of MIO are possible assignments for an SAG or for surface combatants serving with CSGs or ARGs. Tactics include:

1. Area surveillance by aircraft, including ship's helicopters
2. Operations to board, halt, search, turn back, or coerce ships or boats attempting to pass the interception force

3. Those described under CD operations to halt small, covert boats operating from ships at sea
4. Quick reaction tactics involving armed helicopters or armed high-speed boats operating from ships with C2 capabilities
5. Destroying blockade or quarantine runners
6. Missile ship guard to prevent transit by air
7. Self-defense against anti-MIO operations, including threats from gunboats, missile boats, aircraft, and submarines

HUMANITARIAN ASSISTANCE

Humanitarian operations cover a wide variety of missions. They may take place at sea, or anywhere along the littoral where human beings are threatened. Specific missions include:

1. Providing food, water, and medicine to boats
2. Providing supplies to populations ashore
3. Providing technical assistance or labor
4. Orienting, transporting, guarding, and rescuing humanitarian personnel
5. Establishing no-fly zones, cease-fires, or demilitarized zones

NEOS

Combat forces are often assigned to transport or protect the movement of threatened civilians away from zones of impending military operations or other danger. Civil war, genocide, volcanic activity, or other natural or manmade disasters may cause authorities to order NEO operations, or a local commander may, for humanitarian reasons, initiate them on short notice. The planning process is generally compressed, much like that for a raid. In natural disasters, logistics are primary: how many people can be loaded, transported, protected, and fed. In hostile situations, a high degree of combat readiness may be required as well. The commander should plan for the likelihood of attack, including:

1. Demonstrations of force to deter attack
2. Planned responses to terrorist activity
3. Reactive strikes

SPMAGTF

The global geopolitical environment and increased operations in littoral waters require flexibility in packaging and positioning forces. To respond quickly to emerging situations, a CVN may embark an SPMAGTF comprised of several hundred Marine Corps infantry and aviation personnel and helicopters. Due to its small size, the SPMAGTF cannot replace the ARG and MEU (special operations capable) (MEU(SOC)). The SPMAGTF does, however, combine the speed and firepower of the CVBG with a Marine force to make specific Marine Corps capabilities more widely available and more rapidly employable than would otherwise be the case. An embarked SPMAGTF displaces at least two CVW aircraft squadrons, which must be left ashore.

Coordination Missions

Surface combatants are generally highly capable in supporting C3. They are the glue that holds a grouping of naval assets together. They coordinate air, subsurface, and land-based assets, including joint assets (especially air) under some conditions. Coordination missions are:

1. C2 of tactical groupings
2. Subsurface asset coordination
3. Air asset coordination
4. Land-based sensor coordination
5. Space-based sensor coordination
6. Joint task force (JTF) flagship

Coordination can be a primary mission. Some classes, such as the LCC, are designed to support coordination missions. But a destroyer could be assigned as a tactical command ship for a patrol boat squadron, or an LHA could be assigned to coordinate a humanitarian assistance operation.

C2 OF TACTICAL GROUPINGS

Because of their robust communication suites, all surface combatants have some ability to C2 tactical groupings. Flagship-capable units have even more capability. These ships support tactical commanders as follows:

1. LHA/LHD/CG/LCC/AGF — Fleet commanders and CJTF
2. LHA/LHD — Amphibious group commanders and CJTF
3. CV — CVBG/CVBF commanders, joint force air component commander (JFACC), CJTF
4. LHA/LHD/LPH/LPD (flag configured) — ARG/sea control ARG commanders
5. CG — AAW commanders, MAG, and strike SAG commanders
6. DD/DDG/CV — ASW commanders and SAG commanders
7. CV/CG/DD/DDG — ASUW commanders
8. CV/CG — STW commanders.

SUBSURFACE ASSET COORDINATION

Submarines operate with surface ships in four ways:

1. **Area support.** SSNs on independent area operations may be specifically tasked by submarine operating authorities (SUBOPAUTHs) to support surface forces. These tasks are normally executed autonomously with no requirement for the submarine to communicate or cooperate with the supported force. SUBOPAUTH retains tactical control and informs the supported force commander of task status.

2. **Associated support.** Submarines in associated support operate under the tactical control of the SUBOPAUTH who coordinates tasking and movement in response to supported force requirements. The submarine operates independently but communicates with the supported force to exchange intelligence information. It is tasked to coordinate operations with elements of the supported force.

3. **Support with shift of tactical control (defined as "direct support").** The SUBOPAUTH retains tactical command and shifts tactical control of the submarine to an afloat commander. The afloat commander directly controls the tactical movement and actions of assigned submarines in specific waterspace areas designated by the SUBOPAUTH.

4. **Integrated operations (delegation of tactical command).** Upon receiving tactical command from the SUBOPAUTH, the afloat commander assumes responsibility for all operations and safety of assigned submarines, including local waterspace management and prevention of mutual interference.

EAST CHINA SEA (Nov. 19, 2008) The guided-missile cruiser USS *Shiloh* (CG 67) leads a formation of U.S. Navy and Japan Maritime Self-Defense Force ships assigned to the George Washington Carrier Strike Group and the Essex Expeditionary Strike Group during the bilateral exercise ANNUALEX. ANNUALEX is a yearly exercise with the U.S. Navy and the Japan Maritime Self-Defense Force to improve working relations between the two navies. (U.S. Navy Photo by Chief Mass Communication Specialist Ty Swartz/Released)

AIR ASSET COORDINATION

Surface combatants maintain qualified personnel designated as:

1. Air intercept controller (AIC), controlling fighter, attack, and EW aircraft
2. Antisubmarine/antisurface tactical air controller (ASTAC), controlling fixed-wing and helicopter ASW aircraft
3. Air direction controller (ADC), controlling helicopters for missions other than ASW and ASUW. Primary air control ships, such as CGs, can act as AAW commanders, coordinating an air defense or attack wing, including tanker and surveillance aircraft. CGs can also support an embarked tactical commander for that purpose.

Other ships, such as DDGs, have less air control capability but can control several combat air patrols (CAPs) operating in support of the ship or an associated SAG. Cruisers, destroyers, and frigates also have ASTAC capability. A destroyer can control ASW aircraft, both fixed wing and helicopter, in prosecution of several ASW search areas or datums. Aircraft that can be controlled by surface combatants in support of tactical operations are:

1. Fighter aircraft, sea- or land-based, any Service
2. Surveillance aircraft, sea- or land-based, any Service
3. ASW fixed-wing aircraft, carrier- or land-based (Navy)
4. ASW helicopters, carrier- and surface combatant–based (Navy)
5. ASUW helicopters (Navy, Marine Corps, Army, and allied)
6. Allied aircraft of any type or Service. Some surface combatants are configured to launch, control, and recover (if recoverable) UAVs for surveillance, targeting, and OPDEC

LAND-BASED SENSOR COORDINATION

Land-based sensors important to tactical operations include:

1. Intelligence collection and fusion centers
2. Ocean surveillance collection and fusion centers
3. Land-based HFDF centers
4. Land-based over-the-horizon radars

Surface ships can request and use tactical information from all these sources. The commander establishes requirements, coordinates requirements with position and movement, communicates, and minimizes the time required to receive data (time late) to make these sensors tactically useful. Land-based sensors are particularly important to SAGs, MAGs, ARGs, and other tactical groups lacking a CVW. Timely data from land-based sensors can compensate for lack of long-range surveillance assets.

SPACE-BASED SENSOR COORDINATION

Space-based sensors important to tactical operations include:

1. Photographic satellites, providing long-term intelligence, real-time intelligence, targeting, and mapping
2. ES satellites which detect, locate, and identify enemy activities by means of their electromagnetic emissions
3. Radar satellites which detect and locate ships that may not be emitting electromagnetic energy. Surface ships can request and coordinate tactical information from these sources. The planning required for land-based sensors applies equally to space-based assets. Space-based sensors can provide long-range surveillance information to tactical groups without surveillance aircraft.

JTF FLAGSHIP

As joint operations become the rule rather than the exception, surface ships (especially CVs) may host the CJTF and his staff, including the joint force air component commander (JFACC). The JTF has unique C2 requirements that require upgrading the standard CV communications suite configuration. To accommodate JTF needs, C2 spaces normally assigned to the ship's company and the battle group commander must be shared or relinquished.

Command Relationships

The Surface Warfare Commander, SUWC, is one of the warfare commanders subordinate to the Composite Warfare Commander. To conduct SUW the SUWC coordinates use of assets with the other warfare commanders. In a standard Carrier Battle Group setup, the SUWC is normally the CVW or Destroyer Squadron (DESRON) Commander.

The SUWC may detach a *Surface Action Group* (SAG), comprised of two or more ships, to destroy or neutralize enemy surface warships and their cargo carrying capability. The SUWC will also designate a *SAG commander*. Detachment of an SAG creates both an offensive and defensive baseline, and forces the enemy's attention, at least in part, away from the high value units. Thus, friendly forces inside the baseline may concentrate on other targets or warfare areas.

Global Command and Control System–Maritime (GCCS-M)

The Global Command and Control System-Maritime (GCCS-M) previously known as JMCIS (Joint Maritime Control Information System), is the Navy's primary fielded Command and Control System.

The objective of the GCCS-M is to satisfy Fleet requirements through the rapid and efficient development and of display capability. GCCS-M enhances the operational commander's warfighting capability and aids in the decision-making process by receiving, retrieving, and displaying information relative to the current tactical situation. GCCS-M receives, processes, displays, and manages data on the readiness of neutral, friendly, and hostile forces in order to execute the full range of Navy missions (e.g., strategic deterrence, sea control, power projection, etc.) in near-real-time via external communication channels, local area networks (LANs), and direct interfaces with other systems.

The GCCS-M system is comprised of four main variants, Ashore, Afloat, Tactical/Mobile, and Multi-Level Security (MLS) that together provide command and control information to warfighters in all naval environments. GCCS-M provides centrally managed C4I (Command, Control, Computers, Communications, and Information) services to the Fleet allowing both U.S. and allied maritime forces the ability to operate in network-centric warfare operations. GCCS-M is organized to support three different force environments: Afloat, Ashore, and Tactical Mobile. Afloat configurations can be categorized as force-level and unit-level configurations. Ashore configurations of GCCS-M are located in fixed site Tactical command centers designed to provide the Joint Task Force Commander with similar C4I capabilities when forward-deployed ashore. In order to allow for maximum interoperability among GCCS systems at all sites and activities, GCCS-M utilizes common communications media to the maximum extent possible. The Secure Internet Protocol Router Network (SIPRNET), Non-Secure Internet Protocol Router Network (NIPRNET), and the Joint Worldwide Intelligence Communication System (JWICS) provide the necessary Wide Area Network (WAN) connectivity. Operating "system-high" at the Secret and SCI security levels, both networks use the same protocols as the Internet. In addition to the SIPRNET operating at Secret/SCI security levels, GCCS-M supports collaborative planning at the National Command Authority level by providing Top Secret connectivity to a limited number of sites. GCCS-M had been implemented traditionally on high-performance UNIX workstations; until recently only these platforms were powerful enough to run GCCS-M software. However, with the exponential increase in processing capability, migrating GCCS-M to the PC environment is a very practical and logical decision. Now designed for the PC environment, GCCS-M becomes largely hardware independent, meaning that it uses almost all existing hardware platforms. SUW is conducted in four distinct phases. Throughout each phase, the commander must remain prepared for enemy counter-attack.

(1) *Surface, Surveillance, Communications, and Identification Phase (SSC&I)*
(2) *Approach Phase*
(3) *Attack Phase*
(4) *Post-Attack Phase*

SSC&I Phase

The primary objective of the SSC&I phase is to locate, identify, and target potentially hostile contacts. Location, identification, and targeting can be accomplished using either passive or active methods employed by ownship or Over-the-Horizon Targeting (OTH-T) platforms.

PASSIVE METHODS

Passive methods employ receiving sensors only, thus no emissions are vulnerable to detection. For this reason visual identification is considered a passive method. Passive methods typically come from the firing unit's own Electronic Support (ES) information, third-party ES information, a combination (using ES information from multiple units), or external intelligence sources (i.e.; satellites).

ES may provide position information, either through a cross fix or Target Motion Analysis (TMA), and the firing unit may be able to identify and obtain a firing solution on the enemy. Therefore, it is vital that friendly forces monitor known enemy radar and fire control frequencies, as even a momentary detection on ES systems may be enough to finalize the firing solution. Of note however, passive criteria ONLY is no longer an accepted means of targeting and current doctrine no longer supports weapons release based on ES information only. Intelligence information may also come from a satellite or other highly classified sources. However, information from these sources often is provided too late to be of use for immediate targeting. These sources are typically more useful in providing search areas and threat warnings.

ACTIVE METHODS

Active methods employ the use of radiated energy, thus *units are vulnerable to detection*. The returns from this energy are analyzed to determine bearing and range. Thus active methods provide no identification.

Approach Phase

The approach phase is a four-step process.

    STEP 1: *Organize SSG.*

    STEP 2: *Detach SSG to prosecute.*

    STEP 3: *Review target information and pass it to Fire Control Station (FCS).*

    STEP 4: *Maintain emissions control (EMCON) as directed.*

Attack Phase

A successful engagement on a target with a capable anti-ship missile defense requires many missiles from different directions to arrive simultaneously. Considerations of the attack are outlined below.

1. Determination and dissemination of time on top (TOT), simultaneous impact, for desired missiles.
2. Missile inventory.
3. Number of missiles required to neutralize the intended target. (Calculated estimates are available in warfare publications.)
4. The missile selects the first target it sees. Therefore, it is difficult to ensure the high-value unit of an enemy formation is targeted rather than the escorts.
5. Area of Uncertainty (AOU): If targeting information is accurate and timely, the AOU, or section of ocean where the contact could possibly be found, will be small. Conversely, older information will produce very large AOUs covering hundreds of square miles of ocean.
6. Engagement Planning Figure of Merit (EPFOM): When an engagement plan for a contact is created, the weapons systems will calculate the probability of the weapon acquiring the intended

target. This probability is based on the size of the AOU; serious consideration must be given to an engagement plan with a low EPFOM.

Post-Attack Phase

Battle Damage Assessment (BDA) must be conducted after the attack phase. This may be done using ES, radar, visual, or sonar. The BDA will determine a course of action, such as:

1. Attack again
2. Withdraw from the area
3. Detach additional SSG

Over-the-Horizon Targeting (OTH-T)

OTH-T embodies the general concepts of long-range targeting. Typically a forward stationed ship or aircraft transmits information back to the main body for use in fire control systems. Often referred to as pickets these forward stationed platforms employ the following methods to accomplish their objective:

1. Single ship—ESM and TMA
2. Multiple ship—ESM cross fix
3. Ship and Aircraft (AIC)—ESM cross fix, AIC radar/visual

Things to Consider When Conducting OTH-T

1. Do the units have enough angular separation to provide an accurate ES fix?
2. Is the fix achieved through the AIC's radar or visual identification?
3. How vulnerable will the AIC be when it elevates and radiates?
4. Can AIC pop-up, take a few sweeps, and then drop down?
5. Do the Rules of Engagement (ROE) require a visual identification?

Problems With OTH-T

1. **Targeting inaccuracies.** These may occur from the sources listed below.

   (a) Platform position inaccuracies, navigation errors
   (b) Relative platform position inaccuracies, gridlock errors
   (c) Sensor bearing and range inaccuracies, equipment limitations
   (d) AIC communications/link cannot be conducted if the *AIC* is on the deck at a long range. The AIC will have to depart the area and then elevate. This creates a navigation error problem that includes the AIC's own sensor inaccuracies.

2. **Missile flight.** The following factors introduce several possibilities for error.

   (a) Time of flight—A Mach .9 missile requires 7 minutes to fly 63 nm.
   (b) True wind-both cross-range and down-range components will tend to alter the flight path of the missile.
   (c) Target course and speed-3D—knot speed combined with 180 degree alteration in course will result in a large difference from predicted position.
   (d) Missile speed—Air temperature will determine whether missiles will fly faster or slower than designed speed dependent upon air temperature. Missile range is determined by the time of flight. If not compensated for, a missile that is flying faster than normal will overfly the target.

3. **Identification.** The missile has no IFF. It will pick the first contact it sees after its seeker activates.

    (a) You must either set waypoints for the missile or off-set the aim point of the missile so that only the intended target will be inside the seeker pattern. Consequently, a firing ship must know the location of all ships in the area.

    (b) The OTC may have ROE that require a specific EPFOM for the target and for non-targets. (For example a target must have a probability of acquisition of greater than 80 percent and a non-target must have a EPFOM of less than 20 percent.) Neutral or unknown contacts in the area may inhibit shooting.

REVIEW QUESTIONS

1. What are the four phases of SUW?
2. What is the command relationship between the CWC and the SUWC?
3. What tactical missions are tasked to surface combatants?
4. What should be considered when conducting OTH-T?
5. What are some problems with OTH-T?
6. Give some examples of Special Missions.

Undersea Warfare

LEARNING OBJECTIVES

At the end of this lecture the student will be able to:

- Define the four types of anti-submarine operations.
- Define undersea classification terms.
- Explain the difference between urgent and deliberate attacks.
- Explain the purpose of a datum.
- Describe the responsibilities of a Search Attack Unit Commander.

Introduction

Undersea Warfare (USW) is a functional term that better describes what is commonly known as Anti-Submarine Warfare (ASW). However, the terms are not interchangeable. USW includes all military operations and programs being conducted under the surface of the oceans, including but not exclusively ASW. The Submarine Force is the primary USW organization but there are several other communities that contribute to USW. The USW community includes but is not limited to the Integrated Undersea Surveillance System (IUSS) community. Special Operation Forces (SOF), Mine Warfare community, Naval Meteorology and Oceanography Command (METOC), the ASW experts in the surface line and aviation communities, and a wide range of academic institutions, laboratories, systems centers, and other organizations. VADM Edmund Giambastrani, former COMSUBLANT and former Vice Chairman of the Joint Chiefs of Staff, once said, "Undersea warfare is a team sport, a combined arms effort. No one community can perform the Navy's undersea warfare role by itself. The combat power of submarines is magnified by the contributions of our teammates."

The undersea warfare community is working to expand its ability to operate effectively in a challenging medium. The focus is more than just ASW. In recent years, the submarine force has successfully deployed Unmanned Undersea Vehicles (UUV) and Unmanned Aerial Vehicles (UAV). The *Virginia*-class submarine is leading the Fleet in its implementation of Commercial-Off-The-Shelf (COTS) technology, which allows for rapid updating of hardware and software. Historically, systems were designed for use on one platform for one purpose. Using COTS, submarines can continuously improve

USS *Seawolf* (SSN 21)

performance of multiple systems, most notably the sonar system. The Acoustic Rapid COTS Insertion, or ARCI, is the new standard for submarine systems and provides significant improvement in processing performance, while reducing cost because of its flexibility and wide availability. While using the same sonar arrays, ARCI has demonstrated significant improvements in our submarines' ability to detect other submarines. The Advanced SEAL Delivery System (ASDS) is one of the submarine force's most capable weapons, with the ability to effectively deploy a platoon of SEALs deep into the littoral.

Asymmetric Warfare Threats

The Navy has shifted focus from an independent, blue-water, open-ocean force to one capable of handling regional challenges in the littoral region. Numerous independent, analytical studies conclude that 21st-century naval warfare will be marked by the use of asymmetrical methods meant to counter a U.S. Navy whose doctrine and force structure enable robust power projection ashore from the littorals. Asymmetric warfare implies that potential adversaries will use easily acquired weapons systems that exploit perceived weaknesses in our doctrine or capabilities.

A disturbing trend is the increasing proliferation and commercial access to technology developed and deployed by major powers. Thanks to a robust international arms market, a regional power could acquire large numbers of relatively low-cost cruise missiles, simple tactical ballistic missiles, diesel submarines, mines, and information warfare technology for a modest investment. Regional powers also have increasing access to commercial satellites capable of providing the necessary communication, command and control network as well as a detection capability that enables targeting ships at sea. Access to asymmetric systems allows regional peers and future peer competitors to build a creditable anti-access denial area and prevent or delay our Navy's use of the littorals. All have the potential to delay or reduce our Navy's ability to project power from the sea.

Diesel Submarines. Forward presence is one of the Navy's key missions. We cannot count on basing our ships in foreign countries close to deployment areas. Consequently, America has chosen to build only nuclear-powered submarines because of their speed, endurance, and ordnance load. Most countries do not require forward presence and do not have the necessary budget and technology to maintain a modern nuclear submarine force. When outfitted with modern sensors, processing capability, and weapons, these diesel submarines become formidable platforms.

Consider the following historical example. During the 1982 Falkland's War, a single Argentine Type 209 diesel submarine, ARA *SAN LUIS* (S-32) operated in the vicinity of the British task force for over a month. Despite the deployment of five nuclear attacked submarines, 24-hour-per-day airborne antisubmarine warfare operations, and the expenditure of 203 British ASW weapons, the British task force never once detected the Argentine diesel submarine. The ARA *SAN LUIS*, on the other hand, had conducted several attacks on British ships but was unsuccessful because of an improperly maintained fire control system and poor quality torpedoes. The outcome of the war would have been very different had the ARA *SAN LUIS* sunk or severely damaged one of the British small-deck carriers or logistics ships. With good material readiness, diesel subs cause a serious threat to all platforms, including nuclear submarines.

Air-Independent Propulsion Systems

Older diesel submarines must snorkel everyday or every other day to recharge their batteries. This evolution puts the diesel boat in a very vulnerable and detectable situation. To eliminate or reduce this vulnerability, air-independent propulsion (AIP) technology was developed. AIP operates without the need for outside air, is very quiet, and produces little heat. Third-generation AIP systems allow diesel submarines to stay submerged for up to three months. AIP technology does not have the international controls and restrictions that nuclear propulsion technology does. Therefore, countries do not have to get permission to develop and sell this technology. Russia, Germany, Sweden, and France all currently have submarines that feature this technology.

Mines. Like submarines, mines create a psychological fear that comes from not knowing where they are located. They can limit the mobility of any aircraft carrier, surface combatant, amphibious ship, or submarine. They may be simple contact mines (detonated by contact) or influence mines (detonated by a deflection in its magnetic field by the magnetic field of an approaching ship). Many countries now possess the technology to create sophisticated mines with target detection systems. Historically, mine warfare has proven deadly. In World War II, mines accounted for more ships damaged or lost than any other weapon. In Operation Desert Storm, enemy mines damaged the USS *Tripoli* (LPH 10) and USS *Princeton* (CG 59).

Nearly 30 nations manufacture mines; approximately 20 of these nations export their products. Currently, there are about 50 countries that maintain significant sea mine inventories, including Iran, Iraq, North Korea, Cuba, Libya, Russia, and China. Mines can be deployed from almost any surface platform, including fishing boats, patrol craft, and merchant vessels. Aircraft and submarines can also deploy mines.

Submarine Significance

Submarines and anti-submarine warfare remains the primary focus of USW. Although U.S. Navy submarines have been around for over 100 years, it wasn't until World War II that submarines demonstrated their significant contribution to naval dominance. German *unterseeboots*, or U-boats, forced the Allies to commit disproportionately large forces to defend the allied sea lines of communication. Americans learned from German U-boat design, construction, and tactics, and were able to significantly improve submarines and torpedoes in order to successfully deploy them against the Japanese Navy. The results were impressive. U.S. submarines destroyed 1,314 enemy ships totaling 5.3 million tons. This equated to 60 percent of the Japanese tonnage lost from only 2 percent of the American naval force. Their campaign was a critical factor in the industrial collapse of the Japanese war effort. This success did not come without sacrifice. Of the 16,000 U.S. submariners, the Submarine Force lost 375 officers, 3,131 enlisted men, and 52 submarines.

Today, our nuclear submarines are vastly superior to their predecessors. Technological advances in all aspects of warfare have significantly improved submarine design, construction, sensors, communications, and weapons. These improvements only enhance the enduring characteristics of the submarine—stealth, endurance, firepower, mobility, and survivability. Our submarines are now multi-mission platforms that routinely conduct missions that no other platform can do.

Submarines excel at preparing and controlling the littoral battlespace for joint expeditionary forces. Because most regional powers now have substantially improved capabilities to locate, target, and engage non-stealthy platforms in the littorals, the ability of submarines to identify and prosecute threats to surface and other platforms is vital. Through Intelligence, Surveillance, and Reconnaissance (ISR) missions, submarines greatly enhance U.S. policymakers' understanding of enemy and terrorist force dispositions and operational doctrine before the outbreak of hostilities. Likewise, they allow us to decisively engage and destroy key threats at minimal risk. Before a full Carrier Strike Group (CSG) or Expeditionary Strike Group (ESG) with nearly 10,000 Sailors has approached a high threat area, a submarine can have already detected, reported, and neutralized major threats.

Submarine Roles and Missions

The fundamental changes in the U.S. Submarine Force since the end of the Cold War involve major shifts in warfighting concepts and doctrine. These include:

- From deterrence of global war to the support of U.S. national interests in regional crises and conflicts.
- From a primary ASW orientation against nuclear-powered submarines to taking full advantage of the modern submarine's multi-mission capabilities.
- From weapon loadouts of primarily MK 48 torpedoes to Tomahawk Land-Attack missiles or other weapons. This changing operational context has rippled through all elements of U.S. submarine operations, from peace time presence to strategic deterrence.

These transitions in the submarine force are in line with global transitions and the evolving nature of the U.S. Navy. The world order has shifted from a bi-polar superpower alignment to a multi-polar collection of interests. While the likelihood of global conflict may be greatly reduced, there is an increased chance of regional conflict. The composition and operational posture of the U.S. Navy reflects this, having changed from a blue-water emphasis to a littoral emphasis. For the Submarine Force this has meant several changes in roles:

- Prior to the end of the Cold War, ASW was the major role for U.S. attack submarines. Now U.S. submarines are more multi-mission oriented.
- Intelligence gathering has shifted from strategic to tactical reconnaissance.
- The submarine force is synergistically interoperable with other Navy and Joint communities for mission accomplishment, including an emphasis on Carrier Strike Group and Expeditionary Strike Group operations.

- Special Warfare with Special Operating Forces (such as Navy SEALs).
- Intelligence with the Surveillance community.
- Deployment of assets in the irregular warfare component

U.S. nuclear submarines perform numerous critical missions, many in ways that submarines are uniquely able to perform. To understand submarine tactics and strategy, one must first learn the capabilities of these ships.

Intelligence, Surveillance, and Reconnaissance (ISR)

Submarines provide the nation with a crucial intelligence gathering capability that cannot be replicated by other means. While satellites and aircraft are used to garner various types of information, weather, cloud cover, and the locations of collection targets inhibit their operations. In some situations it is difficult to keep a satellite or aircraft in a position to conduct sustained surveillance of a specific area and satellites and aircraft are severely limited in their ability to observe or detect underwater activity.

Employed with multiple sensors and operated with precision, submarines can monitor events in the air, surface, or subsurface littoral domain providing a complete picture of an event across the full spectrum of intelligence disciplines. They are also an intelligence "force-multiplier" by providing real-time intelligence on contacts of interest to other assets. For example, submarines have been used in South America to perform counter-drug operations events in conjunction with surface assets. While performing ISR, a submarine will transmit intelligence to frigates or other surface platforms on drug-running speed boats, for example. This will allow the frigate to pursue the contact of interest, and using their embarked Visit, Board, Search, and Seizure (VBSS) team, interrupt drug trafficking prior to reaching American waters.

Submarines are able to monitor undersea events and phenomena not detectable by any other sensor. Since they are able to conduct extended operations in areas inaccessible to other platforms or systems, submarines can intercept signals of critical importance for monitoring international developments and enable a wide array of military operations. Furthermore, the ability to dwell covertly for extended periods defeats efforts to evade collection or deceive satellites and other sensors.

The unique look angle provided by a submarine operating in the littoral region enables it to intercept high interest signal formats that are inaccessible to reconnaissance satellites or other collection platforms. The intelligence gleaned from submarine operations ranges from highly technical details of military platforms, command and control infrastructure, weapons systems and sensors to unique intelligence of great importance to national policymakers on potential adversaries' strategic and operational intentions. Most importantly, unlike other intelligence systems such as satellites, submarines are also warfighting platforms carrying militarily significant offensive firepower.

Precision Strike

All U.S. attack submarines carry the Tomahawk Land-Attack Missiles (TLAM) capable for torpedo-tube launch, which can be reloaded and fired while submerged. *Virginia*-class and *Los Angeles*-class submarines SSN 719 and later also have 12 additional Tomahawks in a Vertical Launch System (VLS) battery located in the bow. TLAMs provide the capability for long-range, precision strikes with conventional warheads against shore targets.

Typically, submarines provide about 20 percent of the Tomahawk firepower in a CSG. Additionally, because of their stealth, these attack submarines can be positioned to operate alone in environments where the risks would prevent surface and air forces from operating without extensive protective cover. Other advantages of using submarines outfitted with Tomahawks are:

- Air superiority is not required
- Timely flexibility
- No chance of lost aircraft or airmen.

First used in combat in the 1991 Gulf War, the TLAM has proven to be a highly effective weapon. The official Department of Defense report "Conduct of the Persian Gulf War" (1992) states: "The observed accuracy of TLAM, for which unambiguous target imagery is available, met or exceeded the accuracy mission planners predicted." When the war began on the night of 16 January 1991, the opening shots were Tomahawk cruise missiles launched from U.S. Navy surface ships in the Red Sea and the Persian Gulf. The missiles arrived over the heavily defended Iraqi capital of Baghdad at about the same time as U.S. Air Force F-117 stealth attack planes carrying guided bombs. During the six-week air war, F-117 attack planes were the only strike aircraft to operate over Baghdad at night, and TLAMs were the only U.S. weapons to strike the city in daylight during the entire campaign. Conventional aircraft were not used in strikes against Baghdad and certain other Iraqi targets because of the heavy anti-aircraft defenses.

U.S. Navy surface ships and submarines fired 288 land-attack variants of the Tomahawk during the Gulf War. Battleships, cruisers, and destroyers launched 276 of the missiles and 12 were launched from submarines. USS *Louisville* (SSN 724), operating in the Red Sea launched eight missiles, and the USS *Pittsburgh* (SSN 720), operating in the eastern Mediterranean, launched four missiles. These launches demonstrated the ability of the submarine to operate as part of an integrated strike force, with targets and related strike data being communicated to them at sea. In future military operations, submarines will not replace traditional carrier attack aircraft; rather, submarine and surface ship-launched TLAM strikes will be the vanguard of such attacks, destroying early-warning, air-defense, and communications facilities to reduce the threats against manned aircraft. Submarines in particular can reach attack positions without alerting or provoking the intended adversary.

More recently, 381 TLAMs were fired from all platforms during the "Shock and Awe" campaign. A total of 802 would be fired during the initial stages of Operation Iraqi Freedom. Of these, over one-third were fired from attack submarines, including the first salvo, again fired from the *Louisville*.

Special Operations Forces

Submarines have long been used for special operations, carrying Special Operations Forces (SOF), reconnaissance teams, and other special agents on high-risk missions. SEALs, the Sea-Air-Land teams trained for missions behind enemy lines, carry out most special operations by U.S. submarines. Fixed-wing aircraft, helicopter, parachute, or surface craft can insert these Special Forces, but in most scenarios only submarines guarantee covert delivery. Once in the objective area, SEALs can carry out combat search-and-rescue (SAR) operations, reconnaissance, sabotage, diversionary attacks, monitoring of enemy movements or communications, and a host of other clandestine and often high-risk missions. Nuclear-powered submarines are especially well suited for this role because of their high speed, endurance, and stealth. Submarines can recover personnel who parachute from fixed-wing aircraft and rappel down from helicopters into the sea, take them aboard, and subsequently launch them on missions.

Any U.S. submarine can be employed to carry SEALs; however, the Navy has several submarines that have been specially modified to carry swimmers and their equipment more effectively, including the installation of chambers called Dry Deck Shelter (DDS) to house Swimmer Delivery Vehicle (SDV). These submarines retain their full suite of weapons and sensors for operations as attack submarines, and have special fittings and modifications to their air, hydraulic, and other systems to support the DDS. The DDS can be used to transport and launch an SDV or to "lock out" combat swimmers. A DDS can

be installed in about 12 hours and is air-transportable, further increasing special operations flexibility. Several boats in the *Los Angeles*–class, and all ships in the modified *Ohio*–class SSGN class now carry the Advanced SEAL Delivery System (ASDS), a next-generation SOF-capable platform. The SSGN has the ability to also carry the DDS in tandem. USS *Georgia* (Blue) (SSGN 729).

Using these technologies, these forces can exit the DDS and ascend to the surface, bringing with them equipment and rubber rafts, or they can mount an SDV and travel underwater several miles to their objective area. Alternatively, the ASDS acts as a mini-submarine, able to penetrate far deeper into the littoral than the host submarine can. The number of SEALs carried in a submarine for a special operation varies with the mission, duration, target, and other factors. One or more SEAL platoons can be embarked, with additional SEALs to help with mission planning in the submarine and to handle equipment. The SSGN has special berthing and storage spaces for about 50 SEALs, and an onboard, fully integrated Combat Command Center that, when combined have the ability to carry 154 TLAM plus a full torpedo room load-out, which give the SSGN the ability to conduct entire campaigns with little support from other platforms.

Peacetime Engagement/Power Projection

In peacetime, the deployment of submarines in forward areas can demonstrate U.S. interest in the region. The long endurance and high transit speeds of nuclear submarines make them particularly attractive for rapid deployments to forward areas in such circumstances. Once on station the attack submarine can be highly visible; in 1991 U.S. submarines conducted more than 200 port visits to 50 cities around the world.

An opponent's ability to either deny access to or preempt U.S. military presence is seriously limited against submarines. First, it cannot reliably detect their presence. Second, submarines are not threatened by many of the existing or projected access denial weapons. Coastal cruise missiles, tactical ballistic missiles, and weapons of mass destruction pose little or no threat to a submarine. Submarines carry organic mine detection systems allowing them to avoid previously undetected minefields. A credible attack capability against our submarines could be developed only by substantial investment in an attack submarine force comparable to ours. Accordingly, so long as we maintain our investment advantage, submarines will remain one of the most credible, survivable, and potent land attack missile platforms in our arsenal.

Sea Control

The United States is a maritime nation whose trade and military power projection capabilities depend upon assured use of the high seas. Ocean transport provides the vast majority, accounting for over 90 percent of our strategic lift requirements. Keeping those sea lanes open and stopping enemy surface ships and submarines from using the seas is critical to mission success. Attack submarines can control the seas in a variety of scenarios, from general war against a major maritime power to blockades of enemy ports. Attacks against enemy surface ships or submarines can be part of a war of attrition, or such attacks can be directed against specific targets. Submarines are the quintessential sea control platforms, with proven anti-submarine and anti-surface capabilities. This is most easily seen during World War II in the Pacific Theater, as discussed earlier.

Modern U.S. submarines, armed with significantly improved sensors and weapons, are vastly superior to their historical ancestors. They possess unsurpassed abilities to hunt and kill submarines and surface ships on the high seas and in the littorals. U.S. nuclear submarines provide our only assured capability to wrest control of the sea from a determined enemy employing submarines in an area denial role. As a result, today's U.S. Navy, employing a combined arms anti-submarine capability that includes nuclear submarines, is able to sail freely on the world's oceans.

Strategic Deterrence

Strategic deterrence remains a fundamental element of U.S. defense strategy, just as conventional deterrence has become increasingly important since the fall of the Berlin Wall. Nuclear-powered submarines are a principal component of the future U.S. strategic posture. With the number of land-based bombers and intercontinental missiles being reduced, SSBN force will be the only leg of the Strategic Triad still deploying missiles armed with Multiple Independently Targeted Reentry Vehicles (MIRVs). General Colin Powell, U.S. Army, Chairman of the Joint Chiefs of Staff, at a ceremony in April 1992 marking the completion of the 3,000th deterrent patrol, cited the significance of the Navy's SSBN force. General Powell told the submariners: "… no one, *no one*, has done more to prevent conflict. *No one* has made a greater sacrifice for the cause of Peace than you. America's proud missile submarine family. You stand tall among all our heroes of the Cold War."

Because of the invulnerability of nuclear submarines operated in the vast ocean areas, they provide the nation's strategic deterrent more effectively and at less cost than other systems. Our TRIDENT submarines (SSBNs) now carry over 50 percent of our nation's nuclear deterrent using less than 1.5 percent of naval personnel and less than 34 percent of our strategic budget. These boats will form the backbone of the nation's strategic nuclear force well into the 21st century.

Task/Battle Group Operations

Attack submarines are fully integrated into Navy strike group operations. Typically, 1–2 attack submarines are assigned to each strike group. These submarines participate with the strike group in all pre-deployment operational training and exercises. While operating with the battle group, tactical control of the submarines is routinely shifted to commanders embarked on command and control platforms. Likewise, tactical control of NATO submarines is routinely shifted to U.S. commanders.

Mine Warfare

In both covert offensive mining and mine reconnaissance, submarines provide capabilities that no other platform can deliver. The submarine offensive mining capability allows national leaders to precisely place mines for maximum effect without alerting the enemy and with minimal risk. Mine reconnaissance capability from submarine launched Unmanned Undersea Vehicles allows the submarine to covertly detect and report mine danger areas without risk to naval forces. As a result, potential adversaries have fewer clues indicating potential locations of American expeditionary operations and U.S. military planners are better able to exploit the element of surprise.

Submarines carry mines to deny sea areas to enemy surface ships or submarines. The primary mine deployed by submarines today is the Mk 67 Submarine-Launched Mobile Mine (SLMM). The Mk 60 CAPTOR (short for enCAPsulated TORpedo) is the primary ASW mine deployed by other platforms. The SLMM is a torpedo-like weapon that, after being launched by the submarine, can travel several miles to a specific point, where it sinks to the sea floor and activates its mine sensors. It is particularly useful for blockading a harbor or a narrow sea passage, and uses a rudimentary passive sonar system to detect targets, and detonate.

Other USW Participants

As mentioned earlier, no one community can perform the Navy's undersea warfare role by itself. Integrated support between undersea, surface, airborne, and space-based systems are required to ensure that we maintain what the Joint Chiefs of Staff publication Joint Vision 2010 calls "full-dimensional protection." The undersea environment, ranging from the littoral to deep water to polar regions to under-ice, demands a multi-disciplinary approach including intelligence, oceanography, surveillance, employment of multiple sensors and sensor technologies, coordinated multi-platform operations, and underwater weapons. The following are short descriptions of other USW organizations rarely in the spotlight, but whose missions and roles are becoming more critical to the Navy.

Integrated Undersea Surveillance System (IUSS)

The IUSS mission is to provide support for tactical and strategic forces through the detection, classification, tracking, and reporting of subsurface, surface, and air maritime activities. IUSS uses several systems and ships to accomplish their mission.

Sound Surveillance System (SOSUS): SOSUS is a fixed, sea floor acoustic system that provides long-range detection of older and noisier classes of submarines. It has been in use for a number of years, and the SOSUS hydrophone arrays are positioned for the best acoustic intercept of contacts. The hydrophones are located throughout the Atlantic and Pacific Oceans and the Mediterranean and North Seas.

Fixed Distributive System (FDS): FDS is a newer, long-life, low frequency passive acoustic surveillance system for detecting newer, quieter submarines using hydrophones geographically distributed on the sea floor.

Other IUSS systems include the Surveillance Towed Array Sensory System (SURTASS), Advanced Deployable System (ADS), and the SURTASS Low Frequency Active (LFA) system. These systems are deployed on T-AGOS (do we need to name this?) capable surface ships in the Atlantic and Pacific fleets.

Naval Meteorology and Oceanography Command (METOC)

The METOC mission is to provide the warfighter with the right METOC information, in the right format, to give a decisive edge in combat. Interpreting the battlespace environment for our warfighters is the primary function of the naval oceanography community. Navy oceanographers are using every available resource to understand, predict, and portray the natural environment of waves, water, and weather, particularly in the littoral regions.

Other areas that the naval oceanography community is involved with are electronic intelligence and electro-optical collection, submarine launched mobile mines, shallow water ASW, ice cover, bathymetry, and digital nautical charts.

USW Platforms

SURFACE SHIPS

Aircraft Carriers: Aircraft carriers provide a wide range of options to commanders. Their mission is to provide a credible, sustainable, independent forward presence and conventional deterrence in peacetime; to operate as the cornerstone of joint/allied maritime expeditionary forces in times of crisis; and to operate and support aircraft attacks on enemies, protect friendly forces, and engage in sustained independent operations in war. While aircraft carriers have no true stand-alone ASW sensors, they do carry equipment to collect and interpret information provided by other platforms in the strike group, including aircraft, surface ships, and shore facilities.

TICONDEROGA-Class Cruiser (CG-47): These guided missile cruisers are large combat vessels with multiple target response capabilities. These ships are multi-mission (AAW, ASW, ASuW) surface combatants, meaning they are capable of performing Anti-Air Warfare (AAW), Anti-Submarine Warfare (ASW), and Anti-Surface Warfare (ASuW). They are also capable of supporting Carrier Strike Groups or Expeditionary Strike Groups, or may operate independently as flagships of Surface Action Groups (SAG). They carry MK 46 and MK 50 torpedoes and SH-60 helicopters, discussed below.

ARLEIGH BURKE–Class Destroyer (DDG-51): These destroyers help safeguard larger ships in a fleet or battle group. They operate in support of the CSG, ESG, SAG, and many replenishment groups. These are also multi-mission surface combatants. They carry Mk 46 and Mk 50 torpedoes and SH-60 helicopters.

OLIVER HAZARD PERRY–Class Frigate (FFG-7): Frigates fulfill a protective role as ASW combatants, primarily for underway replenishment groups and merchant convoys. FFGs provide some ASW capability, but they have some limitations. Originally designed as cost efficient surface combatants, they lack the multi-mission capability necessary for modern surface combatants faced with multiple, high-technology threats. It should be noted that the designator "FFG" is a bit of a misnomer: most FFG have had the guided missile battery removed as a cost-reducing measure. The Littoral Combat Ship (LCS) will eventually perform its missions. They carry MK 46 and MK 50 torpedoes and SH-60 helicopters.

AIRCRAFT

P-3 Orion: A fixed-wing shore-based maritime patrol aircraft (MPA), it carries radar, Magnetic Anomaly Detection (MAD), sonobuoys, and other means of detection. It can be outfitted with Harpoon missiles, torpedoes, and mines.

SH-60B LAMPS III: This helicopter was developed to specifically to operate from cruisers, destroyers, and frigates. Sensors include radar, MAD, and sonobuoys. The SH-60B can also carry two MK 46 torpedoes. Sonobuoys are processed on the helicopter or linked back to and processed on the surface platform. The data link is a direct pencil beam in the SHF frequency range which is not likely to be detected by enemy ES (Electronic Support). The link also contains voice communication capability and on board navigation data processing. When not operating in the USW mode the SH-60B provides excellent ASuW support by linking its radar data to the surface platform.

SH-60F: This carrier-based variant of the LAMPS III is similar in most respects to the Bravo version with the addition of dipping sonar. Because it has been adapted for close-in USW defense of the aircraft carrier, the radar, MAD, and sonobuoys have been deleted.

SUBMARINES

LOS ANGELES–Class Attack Submarine (SSN-688): This is the most numerous class of nuclear-powered fast attack submarines built by any nation, and will form the backbone of the U.S. attack submarine force well into the 21st century. *LOS ANGELES*-class submarines are fast (greater than 25 knots), have four torpedo tubes, and can carry up to 25 torpedo-tube launched weapons. SSN 719 and later have 12 vertical launch tubes for the Tomahawk cruise missile. The improved *Los Angeles–* class, SSN-751 and later, are referred to as "688i" and are quieter, incorporate an advanced combat system, and are configured for under-ice operations. Their forward diving planes were moved from the sail structure (where they were referred to as fairwater planes) to the bow (now called bow planes), and are retractable. The sail has been strengthened for breaking through ice. All ships carry the Mark 48 ADCAP (ADvanced CAPability) torpedo, and can deploy the Submarine-Launched Mobile Mine (SLMM). They have the ability to use Harpoon missiles, but rarely do.

SEAWOLF–Class Attack Submarine (SSN-21): The intended successor to the 688-class, this submarine has improved machinery, quieter propulsion, and improved weapons systems. They are significantly quieter and faster than the 688, and also feature 8 torpedo tubes. Only three submarines of this design were built. This class introduced the reconfigurable torpedo room, and has the ability to carry up to 50 torpedoes or missiles, or up to 100 mines. The *SEAWOLF* also introduces the advanced AN/BSY-2 combat system, which includes a new larger spherical sonar array, a Wide Aperture Array (WAA), and a new towed-array sonar. Based on rapidly changing political and economic factors, these extremely expensive, very specialized boats were scrapped in favor of the SSN-774. The last ship of the class, USS *Jimmy Carter* (SSN-23) features a 100-foot module inserted for black operations. All ships carry the Mk 48 ADCAP, and can be configured to deploy the SLMM.

VIRGINIA–Class Attack Submarine (SSN-774): This newest class of submarine was designed for multi-mission operations and enhanced operational flexibility. *SEAWOLF–*class quieting has been

Platform	Strengths	Weakness
Submarine	• Quiet • Mk 48 ADCAP torpedo • Long on-station time • Relatively large weapon load • Great acoustic sensors • In the medium	• Communications • Order modifications (slow to respond to changes • Operate slowly for maximum range detections • Tough to use radar
Surface	• Can be less noisy than expected (with Prairie Masker) • Large weapon/sensor load • C3 facilities • Helicopter pad/hangars • Weapon reach • Moderate on-station time • Can operate at all depths with sensor placement	• Short ASW range without a helo • Vulnerable to torpedo attack • Vulnerable during Underway Replenishment (UNREP) • Mk 46/50 torpedoes less effective than ADCAP
Air	• Placement	• Low on-station time (P3 max. about 12 hours) • Small weapons load • No UNREP capability • Communications vital for entire mission • Station keeping/navigation may be difficult

incorporated in a smaller hull without a significant loss in capability. With a focus on the littoral battlespace, the *VIRGINIA* has improved magnetic stealth, sophisticated surveillance capabilities, and Special Warfare enhancements.

VIRGINIA is engineered for maximum flexibility. Its responsiveness to changing missions and threats, and its affordable, integrated electronic systems with commercial-off-the-shelf (COTS) technology ensures state-of-the-art technology introduction throughout the life of the class to avoid unit obsolescence.

The Command, Control, Communications, and Intelligence (C3I) electronics packages also promote maximum flexibility for growth and upgrade. Coupled with the Modular Isolated Deck Structure and open-system architecture, this approach results in a significantly lower cost, yet more effective, command and control structure for fire control, navigation, electronic warfare, and communications connectivity.

The *VIRGINIA's* sonar system is state-of-the-art and has more processing power than other classes of submarines to process and distribute data received from its spherical bow array, high-frequency array suite, dual towed arrays, and flank array suite.

The *VIRGINIA's* sail configuration houses two new photonics masts for improved imaging functions (this replaces traditional periscopes), an improved electronics support measures mast, and multi-mission masts that cover the frequency domain for full-spectrum, high-data-rate communications. The sail is also designed for future installation of a special, mission-configurable mast for enhanced flexibility and warfighting performance.

VIRGINIA is armed with a variety of weapons. It carries the Mk 48 ADCAP torpedo, SLMM, Tomahawk cruise missiles, and Unmanned Undersea Vehicles (UUV). Future ships in the class will have the ability to employ Unmanned Aerial Vehicles. In addition, Tomahawk missiles are carried in vertical launch tubes. The submarine also features an integral Lock-Out-Lock-In chamber for special operations and can host Special Operations Forces' underwater delivery vehicles.

USS *VIRGINIA* (SSN 774)

OHIO-Class Fleet Ballistic Missile Submarine (SSBN 726): The *OHIO*-class fleet ballistic missile submarines provide the sea-based leg of the triad of U.S. strategic forces with the 14 Trident SSBNs each carrying 24 missiles. By virtue of their patrol posture, these submarines are highly survivable; they are also extremely flexible, capable of rapidly retargeting their missiles, should the need arise, using secure and constant at-sea communications links. They are the largest submarines ever built by the United States. *OHIO*-class submarines can carry either the TRIDENT I (C-4) or TRIDENT II (D-5) missiles. In addition, these submarines are fitted with four torpedo tubes for Mk 48 (later ships feature the Mk 48 ADCAP) torpedoes, which, along with countermeasure devices, provide defense against hostile ASW forces. The most important defensive feature of these submarines is their stealth; they are among the quietest nuclear-powered submarines ever built. This inherent feature of the *OHIO*-class, coupled with other characteristics, makes these ships the most survivable element of the nuclear triad.

Two complete crews, designated Blue and Gold, are assigned to each *OHIO*-class submarine. While one crew is at sea operating the submarine, the other is conducting training, attending schools, being evaluated in shore-based simulators, and enjoying leave. By alternating the Blue and Gold crews, with a brief turnover period, the submarines can be kept at sea for considerably longer than with a single crew. *OHIO*-class submarines are specifically designed for extended deterrent patrols.

OHIO-Class Guided Missile Submarine (SSGN 726): The first four ships of the class described above (Ohio, Michigan, Florida, Georgia) have been refueled and converted to powerful submarines as part of an effort to reduce nuclear warhead inventories. These submarines are described in detail earlier in the chapter. Notably, these four vessels can be outfitted with up to 152 TLAMs in their missile tubes, 24 weapons (primarily Mk 48 ADCAP torpedoes) for its four torpedo tubes, and 2 SEAL platoons to use its embarked DDS and/or ASDS.

Platform Strength and Weakness

Torpedoes

Self-propelled, modern torpedoes (especially the ADCAP) feature onboard active and passive sonar capability. Torpedoes may be launched from submarines, surface ships, helicopters and fixed-wing aircraft. The three major torpedoes in the Navy inventory are the Mark 48 heavyweight torpedo (and its ADCAP variant), the Mark 46 lightweight and the Mark 50 advanced lightweight.

The Mk-48 is designed to combat fast, deep-diving nuclear submarines and high performance surface ships. It is carried by all Navy submarines and is the primary ASW weapon. The improved version, Mk-48 Advanced Capability (ADCAP) is replacing the Mk-48s. Mk-48 and Mk-48 ADCAP torpedoes can operate with or without wire guidance and use active and/or passive homing. When launched they execute pre-programmed target search, acquisition, and attack procedures. Both can conduct multiple reattacks if they miss the target. The MK-48 has a range of greater than 5 miles, can dive deeper than 1200 feet, can travel faster than 28 knots, and carry 650 pounds of high explosive.

The Mk-46 torpedo is designed to attack high performance submarines, and is presently identified as the NATO standard. The Mk-46 torpedo is designed to be launched from surface combatant torpedo tubes and fixed- and rotary-wing aircraft. The Mk-46 has a range of 4 miles, can dive deeper than 1200 feet, can travel faster than 28 knots, and carry 98 pounds of PBXN-103 high explosive in a bulk charge.

The Mk-50 is an advanced lightweight torpedo for use against the faster, deeper-diving, and more sophisticated submarines. The Mk-50 can be launched from all ASW aircraft, and from torpedo tubes aboard surface combatant ships. The Mk-50 will eventually replace the MK-46 as the fleet's lightweight torpedo. The MK-50 has a range of greater than 4 miles, can travel faster than 40 knots, and carry approximately 100 pounds of high explosive in a shaped charge.

Mines

Mines are used as a subsurface ASW or ASuW weapon. The Mk-56 ASW mine (the oldest still in use) was developed in 1966. Since that time, more advances in technology have given way to the development of the Mk-60 CAPTOR (short for "enCAPsulated TORpedo"), the Mk-62 and Mk-63 Quickstrike, and the Mk-67 SLMM (Submarine Launched Mobile Mine). Most mines in today's arsenal are aircraft delivered to target.

Submarine Task Group Operations

An attack submarine assigned to task groups is a relatively new way of doing business in the Navy. Submarines and task group commanders have had to learn how to operate together and learn each other's way of doing business. The benefits for the Officer in Tactical Command (OTC) are tremendous. The OTC now has another platform at his disposal and can use the submarine in a variety of roles, such as providing intelligence data or clearing the battlespace prior to the task force's arrival.

Terms

Coordinated submarine/task group operations require a strong command and control (C2) arrangement to avoid conflicting tasking requirements from different commanders for the same platform and to ensure safety of the submarine. To understand the C2 structure, one must understand the terms associated with submarine/task group operations.

The Submarine Operating Authority (SUBOPAUTH) is the naval commander exercising *operational* control (OPCON) of submarines. The submarine force commander or his designated subordinate for a specified area acts as SUBOPAUTH for overall submarine operations.

The Officer in Tactical Command (OTC) is the senior officer present eligible to assume command or the officer to whom he has delegated tactical command. The OTC is usually the commander of a task force or group.

Operational control (OPCON) is the authority delegated to a commander to direct forces assigned so the commander may accomplish specific missions or tasks which are usually limited by function, time, or location; to deploy units concerned; and to retain or assign *tactical* command (TACOM) or control of those units. In naval or joint operations, SUBOPAUTH is designated to exercise OPCON over all assigned submarines. This includes responsibility for overall area-wide waterspace management (WSM, see below) and prevention of mutual interference (PMI, see below) for submarine operations with the commander-in-chief's area of responsibility (AOR). SUBOPAUTH also has responsibility for and control of the submarine broadcast (message traffic).

Tactical command (TACOM) is the authority delegated to a commander to assign tasks to forces under his command for the accomplishment of the mission assigned by higher authority. TACOM includes retention of authority to delegate tactical control (TACON). TACOM of submarines assigned to a joint or naval task force/group can be delegated to an at-sea commander as mutually agreed by the OTC and SUBOPAUTH.

Tactical control is the detailed and usually local direction and control of movements or maneuvers necessary to accomplish missions or tasks assigned. The officer exercising TACON also acts as the weapons control authority for units assigned as consistent with the rules of engagement (ROE). For submarine operations, TACON is confined to the area assigned to each individual submarine, either by the SUBOPAUTH or the OTC, if TACOM was shifted. The submarine is still required to copy the submarine broadcast as directed by the SUBOPAUTH regardless of the communications and reporting requirements levied by the officer exercising TACON.

Waterspace management (WSM) pertains to the allocation of waterspace in terms of anti-submarine (ASW) weapons control to permit the rapid and effective engagement of enemy submarines, while preventing inadvertent attacks on friendly submarines. Essentially, WSM is a set of specifically defined submarine and ASW force operating areas and attack rules. These procedures are implemented by the SUBOPAUTH on behalf of the area or joint operations combatant commander, and should be in place whenever use of ASW weapons by any platform becomes probable. WSM may be applied on any scale (local, regional, theater) depending on the crisis situation, the existing or projected submarine threat, and requirements of the area or combatant commander.

Prevention of Mutual Interference (PMI) procedures are specifically intended to prevent submerged collisions:

- Between friendly submarines
- Between submarines and friendly surface ships' towed bodies
- Between submarines and any other underwater event (e.g., explosive detonations, research submersible operations, oil drillings, etc.)

C2 of Submarines Employed with Surface Forces

Submarines are assigned to tasks groups at one of four levels: Integrated Operations, Direct Support, Associated Support, and Area Operations. The SUBOPAUTH and the OTC should agree upon the level of command and control desired based on mission, C2 capabilities of units assigned, and operational responsiveness required. TACOM of submarines and responsibility for local WSM and PMI can be delegated to the OTC. But in all cases, SUBOPAUTH retains OPCON of all assigned submarines.

In Integrated Operations, a submarine will provide support to a specific Task Group or Task Force, and the OTC to exercise TACOM *and* TACON of the assigned submarine. OTC assumes the responsibility for all operations and safety of assigned submarines, including local WSM and PMI for their designated area. SUBOPAUTH retains OPCON.

In Direct Support operations, the SUBOPAUTH retains TACOM and OPCON, and shifts TACON to an afloat commander (not necessarily the submarine). The OTC or designated subordinate commander directly controls the tactical movement and actions of assigned submarines in specific waterspace areas designated by the SUBOPAUTH.

In Associated Support operations, submarines operate under the TACON of the SUBOPAUTH, who coordinates specific unit tasking and movement in response to requirements of the OTC or designated subordinate commander. In this role, the submarine communicates with the supported force for exchange of intelligence and is tasked to coordinate operations with elements or units of that force.

In Area Operations, submarines are operating independently of the task force but may be specifically tasked by the SUBOPAUTH in roles that support the particular objectives of a surface force. These tasks are normally executed autonomously with no requirement for the submarine to communicate or cooperate with the supported force. The SUBOPAUTH informs the task force/group commander concerning status and completion of submarine tasks assigned.

• Detect subsurface contacts in enough time to be able to divert Battle group around threat	• MPA • SSN in direct support • Tail ships
• Sanitize base course (main body) • Localize, track, and classify all contacts • If ROE permits, attack	• Tail ship, P-3, S-3, SH-60, SH-3
• Same as Middle Zone except the time for engagement is shorter. Attack from enemy torpedoes is considered imminent. Maximum self-defense measures must be implemented.	• Main body units (IIVU) and escorts. All air except P-3 and S-3. Tail- and hull-mounted sonar surface assets.

Submarine Search and Attack

Stealth is one of the submarine's fundamental characteristics that make the submarine such a formidable platform. Most missions and tactics are classified. RADM Fages, Director, Submarine Warfare Division (N87), wrote, "Stealth allows the imposition of force at the time and place of one's choosing. It creates uncertainty in the mind of an adversary, and it imposes financial and operational costs to counter the submarine that may be lurking in his littorals."

Attack Philosophy.

The successful commanding officer (CO) will maintain stealth and the element of surprise until he is ready for the kill. Once his weapons are away, the submarine CO knows that his presence has been compromised. Therefore, the CO will delay alerting his enemy of his presence for as long as possible. When hunting enemy submarines or other warships, the submarine commanding officer will accept risks and reduce the margin of safety normally maintained in peacetime.

To maintain a high state of readiness, submarines constantly train, using drills, watch section training, or divisional training. In addition, the submarine strives to maintain peak system performance at all times. This is accomplished through frequent and rigorous weapons systems alignment checks, fire control and sensor preventive maintenance, and self-radiated noise monitoring.

There are three phases the submarine follows when destroying an enemy warship:

1. **Contact Phase.** This is the period of time from initial detection of a contact until determination of tentative classification and initial range assessment. Detection of a contact is accomplished by active or passive sonar (submerged) or by visual means (surfaced or at periscope depth). All initial contacts are treated as potential threats unless proven otherwise.

2. **Approach Phase.** When the contact is classified as a threat, the fire control team will develop a target motion solution to obtain adequate firing information. The submarine may close the contact as necessary to get better information.

3. **Attack Phase.** This phase begins when the CO decides to put ordnance in the water and ends when the target is sunk or the threat is neutralized.

TASK FORCE USW OPERATIONS

The USWC (Undersea Warfare Commander) may be a Commanding Officer of a ship, or embarked staff (for example, COMDESRON 12). The USWC reports to the Composite Warfare Commander (CWC) all matters concerning USW operations. The USWC is in charge of all weapons and sensors on all USW ships in the Battle Group.

A Search and Attack Unit (SAU) is one or more ships separately organized or detached from a formation as a tactical unit to search for and destroy submarines. Fixed-wing aircraft or helicopters may augment the SAU. The SAU commander is responsible for:

- Deciding best approach to datum (discussed later)
- Promulgating search and attack plans (from ATP 1(C)–Vol. I)
- Designating attacking unit
- Keeping USWC informed

There are several factors to consider when organizing a task force that will operate in an USW environment. Among them:

Mission.

Threat: diesel or nuclear submarine. If the threat is diesel, the sub's task force will normally use active sonar. If the threat is diesel, the sonar will be passive. This distinction is made because diesel submarines are significantly quieter and harder to find than nuclear submarines, so in order to avoid wasting time, surface ships will employ active sonar to accelerate contact detection. (Note: a submarine conducting ASW will never use active sonar because it exposes their position.) What weapons do they carry?

Capabilities and limitations of own forces

Area of Operations (Open Ocean? Choke point?)

Undersea Warfare operations are classified as *protective* and *offensive actions*. USW protective actions include escort/screen duties, support operations, or harbor defense. In offensive USW, USW units search out, locate, and attack enemy submarines. Protective or defensive USW normally occurs when the presence of an enemy submarine is unknown. When that presence becomes known, or when an enemy sub is spotted, offensive USW is used to attack or shoulder.

A submarine is said to be operating within a specific zone relative to your position or the position of the high value unit. The Outer, Middle, and Inner Zones make up the USW geographic area. Objectives and Platforms associated with these zones are tabulated in the above table.

Zone Objectives Platforms

PROTECTIVE USW ACTIONS

The purpose of protective USW actions is to protect the main body or a convoy. These techniques were developed to aid the USWC in determining danger positions by plotting the submarine weapons capability and intercept problem.

Torpedo Danger Zone (TDZ): This is a 10,000-yard circle plotted around the main body or convoy. Taking the formation speed into consideration, a submarine determines effective conventional torpedo range. The faster the formation speed, the more skewed the TDZ is plotted ahead.

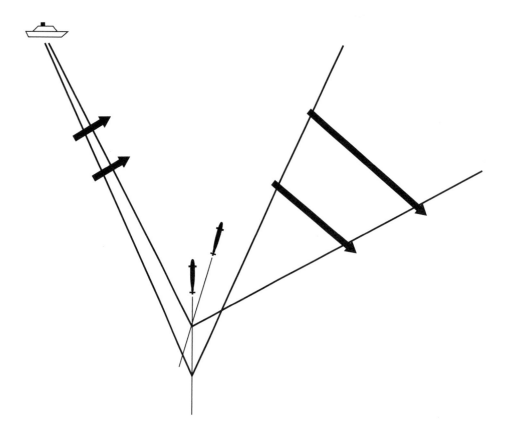

Limiting Lines of Approach (LLOA): The limiting lines of approach determine whether a submarine, knowing the course and speed of a force, can reach the TDZ in a direct line. These lines are plotted tangent to the TDZ circle. As the formation speed increases, the LLOA will tend to narrow. Conversely, as the formation speed decreases the LLOA tend to widen making USW defense more difficult.

Zigzag Patterns: Often formations and convoys will alter course at regular intervals around a base course. This zigzagging makes it harder for a submarine to obtain a fire control solution.

OFFENSIVE USW ACTIONS

Terms. Offensive USW actions attempt to neutralize the sub threat when the enemy sub presence is identified.

Datum: The last known position of a submarine. This position is plotted and is the starting point of a prosecution. Obviously, time is of the essence. Navigation and sensor inaccuracies must be considered when plotting datum error. This is the area of ocean that an enemy sub might actually be operating in.

Furthest on Circles (FOC): These are circles that expand from datum based on time and estimated speed of the sub. It is very similar to plotting dead reckoning, only this is 360 degrees around datum.

Torpedo Danger Area (TDA): This is approximately a 8,000-yard extension plotted beyond the most recent FOC, where an entering prosecuting ship must take due precautions to avoid a torpedo attack.

Cone of Courses (COC): The range of course and speed combinations that a sub may use to intercept the main body/convoy TDZ from datum. This provides the initial search area for the prosecuting assets en route in an offensive role.

Approaches to Datum. There are several different techniques used by prosecuting assets when approaching datum in relation to the position of the main body /convoy.

DIRECT APPROACH: Used by attacking assets when the position of datum (or sub) is less than 6 nm beyond TDZ.

OFFSET APPROACH: Used by attacking assets when the distance to datum (or sub) is greater than the 6 nm beyond TDZ but within +/-30 degrees of main body/convoy track. The offset is 10 to 30 degrees from a direct course to datum offset toward the formation course.

INTERCEPT APPROACH: Used by attacking assets when the distance to datum (or sub) is greater than the 6 nm beyond TDZ and also greater than 30 degrees of the main body/convoy track. Traditionally, MoBoards have been used, but technology advances have shifted this to computer-based solutions.

Search and Attack Method. This is a generalized approach for ASW employed by air, surface, and subsurface platforms.

STEP 1: *DETECTION*—May come first in the form of a passive detection on a surface ship's Tactical Towed Array Sonar System (TACTASS). First detection may also come from a sonobuoy field laid by aircraft, normally a P-3C. Seldom, the initial detection is from a helicopter, as they are typically response platforms and not search platforms due to the low number of sonobuoys onboard, and relative short on-station time. Initial detection is most likely using surface ship active sonar or submarine passive sonar. Submarines rarely will use active sonar as it exposes position, and negates the submarine's stealth and tactical advantage. Submarines will use spherical array passive broadband (SAPBB) as a primary means of platform detection. SAPBB will detect general disturbances in the sound spectrum, which will then be analyzed for classification. Broadband detection systems are generally used for finding *all* contacts. Submarines also feature advanced narrowband detection systems, which will detect specific frequencies or tonals, and are typically used for finding *specific* contacts. For example, when coming to periscope depth, submarines will generally rely on SAPBB to detect all contacts in the area and determine a safe course. When searching for a Chinese *Kilo* or Russian *Akula*, narrowband will be used to find specific tonals emitted by pumps or turbines on the enemy contact.

STEP 2: *CLASSIFICATION*—The decision as to whether or not a contact is a submarine. There are four classifications:

CERTSUB—A contact that has been sighted and positively identified as a submarine is classified CERTSUB (certain submarine) or a weapon is fired. For example, a lookout has seen a periscope in water.

PROBSUB—A contact that displays strong cumulative evidence of being a submarine is classified PROBSUB (probable submarine). This classification is based on the evaluation of data from one or more sensors (radar, sonar, MAD, sonobuoy, etc.).

POSSUB—The classification POSSUB (possible submarine) is applied to a contact on which available information indicates the likely presence of a submarine, but on which there is insufficient evidence to justify a higher classification. The classification POSSUB must always be accompanied by an assessment of the confidence level as follows:

Low confidence—A contact that cannot be regarded as a non-submarine and which requires further investigation.

High confidence—A contact which, from the evidence available, is firmly believed to be a submarine but does not meet the criteria for PROBSUB. A single sensor contact usually meets these criteria.

NONSUB—The evaluator is entirely satisfied that the contact is not a submarine.

STEP 3: *LOCALIZATION*—The process by which one of the contact's locations is better defined is called localization. Submarine spherical array is able to positively identify a bearing to a possible contact fairly easily because of the tear-drop-shaped "beam" characteristics for each hydrophone, typically a few degrees horizontally and a few degrees vertically. Towed arrays, however, because of the elongated design, have hydrophones listening in 360-degree, conical beams. This will result in the array sensing a contact along one of two relative bearings. This is known as bearing ambiguity, which can be resolved either by geography, intelligence, other sensors, or Target Motion Analysis (TMA). The TMA method to resolve bearing ambiguity requires turning the ship to a new heading and finding the new line of relative bearing on the tail. The bearing ambiguity is then resolved, because the new relative bearing will intersect only one of the first two bearings.

STEP 4: *TRACKING*—The contact is then tracked using either active or passive means. If possible, passive tracking is the first choice so as not to alert the target that he is being tracked. There are pros/cons of active and passive sensors, and their use against nuclear and diesel is discussed earlier in this chapter. The objective in tracking is to determine course, speed, and depth to obtain a targeting solution or attack criteria.

STEP 5: *ATTACK*—The placement of a weapon on the target. Typically only one weapon will be fired at one time, and typically from the disengaged side to minimize risk of counter-detection based on torpedo launch sound transient (for example, for a contact on a submarine's port side, the submarine will engage with a torpedo loaded in one of the starboard torpedo tubes). Multiple torpedoes are rarely used because they will tend to interfere with each other. The platform of choice is an aircraft so as not to place a surface ship or submarine within the enemy's weapons range.

URGENT ATTACK: An attack conducted urgently with less regard to exact submarine location and high emphasis on timeliness. Even without exact location of the submarine, placing a torpedo anywhere in the general area of a submarine gives that boat a problem that must be dealt with. The enemy submarine will need to conduct evasive procedures to ensure safety of ship. Often, an urgent attack may be a "snapshot." Snapshots are torpedoes fired as a response to an inbound hostile weapon, normally just down the bearing of the weapon. A stealthy submarine can frequently engage a surface contact without being detected, and the surface ship's first hint of a submarine in the area is the inbound torpedo. This is a purely defensive measure, intended to disrupt the enemy submarine's firing plan while our own ship is evading the hostile weapon.

Example of towed array bearing ambiguity resolution.

DELIBERATE ATTACK: As the name implies, precise location and accuracy are of essence rather than timeliness. This tends to occur when the main body is well protected or has been diverted and is in no immediate danger from a torpedo. Deliberate attacks are typically performed when a platform has been tracking the enemy for some time, and has a very good firing solution, including range, speed, and bearing.

REVIEW QUESTIONS

1. What is USW?
2. Why is USW important?
3. What is asymmetric warfare?
4. How does asymmetric warfare affect modern warfare?

5. Why are diesel submarines a threat to CSG?

6. How are modern submarines different from their predecessors?

7. Identify how submarine roles and missions have changed since the end of the Cold War. Why?

8. What are some strengths and weaknesses of the various USW platforms?

9. What is waterspace management?

10. What is the difference between SUBOPAUTH and OTC?

11. What is the SAU commander responsible for?

12. Identify and describe two techniques used by surface combatants to hinder a submarine's ability to obtain a firing solution.

13. What is meant by protective and offensive USW measures?

14. What are some factors and considerations for the task force in USW?

15. What are the three USW geographic divisions?

16. What are the classification criteria used for submerged contacts?

CHAPTER 10

Air Warfare

LEARNING OBJECTIVES

At the end of this lecture the student will be able to:

- Comprehend how Air Warfare doctrines contribute to the basic sea control and power projection missions of the naval Service.
- More specifically: Define AW. Describe offensive and defensive measures of AW.
- Describe the concept of Defense in Depth.
- Describe the phases of AW.
- Describe the various platforms and weapons involved in AW.
- Describe the factors and considerations that affect AW planning.

Additional Required Reading

None.

Air Warfare (AW) is defined as the actions required to destroy or reduce the enemy air and missile threat to an acceptable level. It includes both offensive and defensive measures.

(1) *Offensive measures* include strikes against ships, air bases, and missile sites. Often these offensive measures fall under the purview of the Strike Warfare Commander. The AWC and STWC must therefore work together closely under the direction of the CWC. Offensive measures are covered in some detail in the chapter on "Strike Warfare."

(2) *Defensive measures* include the use of interceptors, surface-to-air missiles (SAM) and air-to-air (AAM) missiles, guns, electronic countermeasures, cover, concealment, dispersion, and mobility to protect or defend against an offensive threat. Successful conduct of defensive AW involves the proper employment of the *Defense in Depth* concept. This concept includes the coordination of detection equipment, weapons systems, and communications and depends on the tactics employed, the disposition of the force, the operating efficiency of equipment, and the capability of personnel.

Defense in Depth

The Defense in Depth concept includes an Air Warfare area encompassing the total region to be protected from enemy air attack. The area is divided into three areas: *Surveillance Area, Detection Area, and Vital Area.* The surveillance area extends from the center of the area to be protected to the maximum detection range of the battle group. This detection range depends upon the location of the ships and aircraft in company and the type of sensors being employed. For this reason the surveillance area should not be thought of as a circle but rather an area that is constantly changing as units move around using different sensors.

Classification, Identification, and Engagement Area (CIEA)

The CIEA is that region of coverage in which AW weapons may be employed against an air threat. The actual size and shape of this area depends on the geographical position of different AW assets and their own weapons capabilities. It may be further divided into zones:

- Fighter Engagement Zone (FEZ). Engage enemy with air assets.
- Missile Engagement Zone (MEZ). Engage enemy with SAM.
- Close-in Engagement Zone (CEZ). Engage enemy with guns, CIWS, and BPDMS.
- Joint Engagement Zone (JEZ). Engage enemy with multiple air defense weapon systems (SAM and fighters) of one or more Service components simultaneously.

Successful JEZ operations require that all air defense systems are capable of discerning between enemy, neutral, and friendly air vehicles in a highly complex environment. If this requirement cannot be met, separate zones for fighter and missile engagement zones should be established.

In the CIEA, the AWC must identify all air contacts and classify them as friendly, unknown, or hostile. There are a variety of methods for identifying air contacts; it is vital that AW watchstanders be familiar with all of them and be able to use the data they learn to make instantaneous decisions whether to engage or not engage a target. No single ID method can provide a positive classification and identification; they must be used together to help the watchstander make an analysis of his target. These methods include:

- **Identification Friend or Foe (IFF)**—The use of radar transponder codes to determine whether or not an aircraft is friendly. There are four modes of IFF. The first three are used by civilian and military aircraft to identify themselves to air traffic controllers. All U.S. and NATO aircraft are capable of transmitting Mode 4 signals, which are encrypted and identify the aircraft as a definite "friend."
- **Profile**—By examining an aircraft's flight profile, a radar operator can determine its speed, position, direction of flight, point of origin, and (depending on the type of radar) its altitude. This information can be used to make an educated guess about a type of aircraft. For instance, an aircraft flying at 35,000 feet, 500 knots, on a commercial air route appears to be a commercial airliner. However, an aircraft traveling at Mach 2 is almost certainly a military aircraft.
- **Indications & Warning (I&W)**—Intelligence sources can provide information about a specific aircraft or a possible threat than can correlate with a newly detected aircraft.
- **Electronic Sensing (ES)**—ES equipment can detect and classify emitters from unknown aircraft of missiles and provide analysis that helps to identify the target.
- **Visual Identification (VID)**—Personnel on ships or in a friendly airplane can identify unknown airplanes/missiles by sight and pass that information to AW.

Vital Area

The vital area contains the unit or units crucial to mission success and is considered to extend from those units to the maximum weapons release range of the enemy's weapons.

There are two major phases of an AW operation: *Surveillance/Detection* and *Engagement.*

Phase 1: Surveillance/Detection

Phase 1 is a continuous operation. At no point is the intensity of phase one lowered or terminated, despite the threat situation. Working forward, from the sea in the littoral environment, time is of the essence. Low, slow flyers can be the greatest challenge to our sensors. Working close to land it is imperative to identify an unknown air contact and have a fire control solution prior to the contact reaching weapons release range. *It is much easier to shoot the archer than the arrow.*

AIRBORNE EARLY WARNING AIRCRAFT

Timely warning of airborne threats is enhanced by employing *airborne early warning (AEW) aircraft.* AEW aircraft are equipped with an airborne tactical data system (ATDS) that is capable of linking with the shipboard NTDS. They are very good at detecting low-flying aircraft and missiles, can conduct air controlled intercepts, and act as a communication relay between the force and distant stations.

PICKET UNITS

Picket units are stationed in an air corridor through which all friendly aircraft must pass when returning so that they may be "deloused" of any shadowing enemy aircraft and given safe passage to recovery. Any combatant ship or AEW aircraft capable of air intercept control may be assigned picket duties in addition to their surveillance function. AW pickets are explained in greater depth below:

(1) *RADCAP.* Fighter aircraft assigned as airborne radar picket.
(2) *PIRAZ/SSS (positive identification radar advisory zone/strike support ship)* is an NTDS equipped missile ship that can simultaneously track, analyze, and display multiple air contacts and broadcast them over link 11 or 16. Ships assigned PIRAZ/SSS duties are stationed well in advance of the main forces in the direction of the threat, often in close proximity to enemy territory. They are capable of positively identifying all friendly aircraft, serving as navigational checkpoints for strike aircraft, controlling air intercepts, coordinating search and rescue operations, maintaining plots of the air picture in CIC, and performing other duties as required to ensure protection of the force.

Phase 2: Raid Engagement

Phase 2 is a specific action against a specific threat that ends when the threat is eliminated.

COMBAT AIR PATROL (CAP)/DEFENSIVE COUNTER-AIR (DCA)

Fighter aircraft under the control of PIRAZ or other picket units are a force's first line of defense against air attack. Fighters are stationed beyond shipboard missile range and are vectored to the most favorable attack position by air intercept controllers (AIC) aboard ship. An E-2C and a group of F/A-18s are teamed up to take advantage of the Hawkeye's powerful radar and the fighter's excellent intercept capabilities. Fighters on CAP station usually fly in fuel-conserving "racetrack" patterns. When vectored, fighters may intercept detected threats at ranges of hundreds of miles from the carrier.

	F/A-18 A/B/C/D	F/A-18 E/F
Manufacturer	Prime: McDonnell Douglas; Major Subcontractor: Northrop.	McDonnell Douglas
Crew	A and C models: One B and D models: Two	E model: One F model: Two
Weights	51,900 pounds	66,000 pounds
Speed	Mach 1.7+	Mach 1.8+
Ceiling	50,000+ feet	50,000+ feet
Combat Radius	1,089 nautical miles (1252.4 miles/2,003 km), clean plus two AIM-9s Ferry: 1,546 nautical miles, two AIM-9s plus three 330 gallon tanks.	Combat: 1,275 nautical miles (2,346 kilometers), clean plus two AIM-9s Ferry: 1,660 nautical miles, two AIM-9s, three 480 gallon tanks retained.
Armament	One M61A1/A2 Vulcan 20mm cannon; AIM 9 Sidewinder, AIM 7 Sparrow, AIM-120 AMRAAM, Harpoon, Harm, SLAM, SLAM-ER, Maverick missiles; Joint Stand-Off Weapon (JSOW); Joint Direct Attack Munition (JDAM); various general purpose bombs, mines, and rockets.	One M61A1/A2 Vulcan 20mm cannon; AIM 9 Sidewinder, AIM-9X (projected), AIM 7 Sparrow, AIM-120 AMRAAM, Harpoon, Harm, SLAM, SLAM-ER (projected), Maverick missiles; Joint Stand-Off Weapon (JSOW); Joint Direct Attack Munition (JDAM); Data Link Pod; Paveway Laser-Guided Bomb; various general purpose bombs, mines, and rockets.

AUTONOMOUS CAP/DCA

A CAP unit stationed at large distances (300 nm) from the main body serves a dual purpose of extending the force's surveillance perimeter, usually along the threat axis, and are free to engage enemy threats based on specific Rules of Engagement (ROE).

AIRCRAFT

Aircraft are an essential part of the AW Defense in Depth concept. Serving both as surveillance platforms and attack platforms the versatile aircraft of the U.S. fleet can meet the challenges of the intensive AW environment head on.

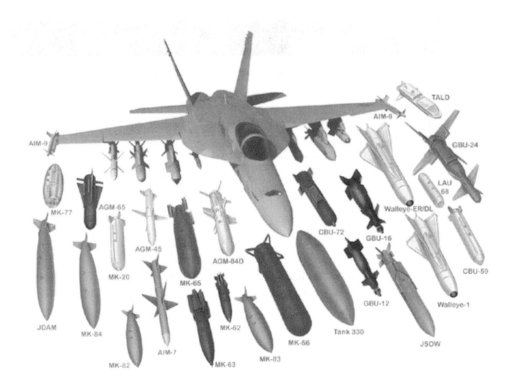

The F/A-18 E/F Super Hornet is the world's most advanced high-performance strike fighter. Designed to operate from aircraft carriers and land bases, the versatile Super Hornet can undertake virtually any combat mission. It provides adverse weather day and night precision weapons delivery. The F/A-18 E/F Super Hornet is leading Naval Aviation into the 21st century.

Missile Designations

Missiles as with aircraft, have both a name and a military designation.

For example, RIM-2E is a ship-launched guided missile designed to intercept air targets. The E indicates that it is the fifth upgraded modification to the design.

Surface-to-air missiles (SAMs) are also categorized according to their range: short, medium, or long.

The most common missiles used by the U.S. Navy are the *Sparrow, Sidewinder, Standard Missiles (SM series), Sea Sparrow,* and *RAM.*

Sparrow (AIM–7) is a medium-range semi-active radar guided weapon. It has a speed of about Mach 2.5 and carries an 88 pound high–explosive warhead. It is an all aspect missile (i.e., it can attack an enemy aircraft from any direction) and it can operate in all types of weather. Platforms: fighter aircraft *(F/A–18)*

Sidewinder (AIM–9) is the most widely used air-to-air missile. As a heat–seeking missile, it is totally passive and requires very little equipment on the launching platform, making it a highly versatile fire and forget missile and capable of being placed on almost any aircraft. The latest models can be used against a target from any aspect. The Sidewinder has a speed of Mach 2+. Its effectiveness is reduced in bad weather. Platforms: fighter aircraft *(F/A–18),* attack aircraft *(F/A–18,* AV–8), helicopters (AR–I), others (P–3)

Standard missiles (SM) is a generic term designating one of three missiles all sharing a common airframe. All have semi–active radar homing guidance with a high explosive, or continuous rod warhead. All have a speed in excess of Mach 2. It has limited use in the surface–to–surface mode against enemy surface threats.

Launching Platform	Mission	Type
A – Air launched	D – Decoy	M – Guided missile
B – Multiple capable	E – Special electronic	R – Rocket
C – Coffin stored horizontally	G – Surface attack designed to destroy enemy land or sea targets	N – Probe
F – Individual carried and launched	I – Intercept-Aerial designed to intercept aerial targets in defensive roles	
M – Mobile launched	Q – Drone	
P – Soft pad launched	T – Training	
U – Underwater launched	U – Underwater attack	
R – Ship launched	W – Weather	

Seasparrow (RIM–7M). The NATO Seasparrow is the Sparrow (AIM–7) missile modified for ship-board use in the basic point defense missile system (BPDMS). Although Seasparrow is designed to protect individual ships against aircraft and missile attacks, it also has limited use against surface targets. Platforms: aircraft carriers, destroyers, frigates, command ships (LCC), amphibious ships (LHA, LPH).

RAM (Rolling Airframe Missile) (RIM–116A) is a modified Sidewinder for use by surface ships as a Basic Point Defense Missile System (BPDMS) against incoming aircraft and cruise missiles. This relatively new missile is just being introduced into the fleet.

Guns

Guns are a ship's final defense against air attack. Most guns that the U.S. Navy uses today are dual–purpose (DP) guns that can be used both against air and surface targets.

The 5"/54 Caliber automatic rapid–fire DP gun is carried by virtually all post–World War II destroyers and cruisers. The 70–pound shell has an maximum range of about 12 nm. The lightweight, unmanned Mk 45 gunmount is the most common variant of this gun in the fleet today. An older manned 5" gunmount, the Mk 42, is an older version no longer used on our ships.

The 76–mm/62 Caliber Oto Melara Mk 75 rapid fire DP gun coupled with the Mk 92 GFCS is a highly accurate, lightweight gun system standard on all FFG–7, and most Coast Guard Cutters. It has a firing rate of 85 rounds per minute and a range of about 9 miles.

The *20mm Phalanx Close–In Weapon system (CIWS)* is designed to be a ship's last ditch defense against anti–ship missiles. The system includes its own search and tracking radar, a fire control system, and the 20–mm, six–barrel Vulcan cannon capable of firing 3,000 rounds a minute at an effective range of up to 1 nautical mile.

The Standard Missile Family	
SM-1 Block 5 MR — RIM-66B	SM-2 Block IV — RIM-156A
SM-1 Block 6B — RIM-66E	SM-2 Block IVA — RIM-156B
SM-2 Block 2 MR — RIM-66C/D	SM-3 Block I — RIM-161
SM-2 Block 3 MR — RIM-66G/H/J	SM-3 Block IA — RIM-161A
SM-2 Block 3A MR — RIM-66L	SM-3 Block IB — N/A
SM-2 Block 3B — RIM-66M	

Air Warfare Planning

Disposition (position of units) of a modern naval force must be flexible enough to cope with multiple, perhaps simultaneous, threats. Not only must the degree of a particular threat be taken into account, but also the force's surveillance and engagement requirements. The best formation for surveillance may not be the best for defense against air attacks. A good air defense formation may present problems in defending against enemy surface and subsurface threats. The enemy air threat may include a variety of air and surface launched missiles. Primary defense is the neutralization of launching platforms. Additionally, operating in the littoral environment presents new concerns about reaction time. Thus, flexibility is of prime importance.

Factors to consider when determining an AW disposition include:

- Coverage required (type, geographical area, threat sector/threat axis).
- Ships and aircraft available for picket duty.
- Communications, capabilities, and limitations.
- CAP capabilities.
- EW capabilities.
- Existing environmental conditions.

The last factor is very important, and often critical in AW. Environmental conditions can greatly affect the performance of sensors and communications, not only in the AW context, but in all other areas of naval warfare.

Command and Control in Antiair Warfare

SECTOR AAW COMMANDER

If the tactical situation dictates that the AAW area should be divided into sectors, a sector AAW commander (SAAWC) may be designated. Within his sector, an SAAWC will be subject to the overriding authority (VETO) of the AAWC, but is otherwise vested with most of the prerogatives and responsibilities of the AAWC.

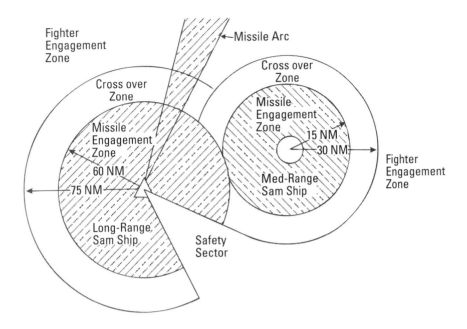

LOCAL AAW COORDINATOR

When two or more ships are in company and within UHF range, a local AAW coordinator (LAAWC) will be designated. The AAWC of a force will normally act as the LAAWC for his own local group.

Positioning Forces

When forces are properly positioned, the concept of Defense in Depth can be properly employed. Within the areas of AW, various lines of defense are established in order to protect the main body or high value units. The following guidelines are utilized to position forces in accordance with the considerations above.

Missile Engagement Zone (MEZ)

Standard SAM MEZ. An MEZ is a zone around a ship or force in which targets are to be engaged by ship's missiles, in accordance with weapon control status/orders.

Silent SAM MEZ. When covert, long- or medium-range, SAM-equipped ships are stationed upthreat, a Silent SAM MEZ may be established; this has significant differences from the MEZ described above. The key factors for ships operating a Silent SAM MEZ are:

1. Ships remain covert, receiving the air picture via data links.
2. Ships' weapon control status is automatically WEAPONS FREE.
3. The OPTASK AAW will contain details of MEZ size, position, and timing.
4. No safety sectors are established in a Silent SAM MEZ.
5. No friendly aircraft are allowed in the Silent SAM MEZ, except for AEW, ASW, and ASUW aircraft that are required to operate in the Silent SAM MEZ, providing the following criteria are met:
   (a) SAM ships must be alerted to the mission, the aircraft must be continuously tracked, and their position transmitted on data link.
   (b) Gridlock between the data-link reporting unit and SAM ship is excellent.

Crossover Zone

A crossover zone normally extends 15 nm beyond the MEZ, but the width may be varied by the AAWC and is usually based on the speed of the assumed threat.

Fighter Engagement Zone

Fighters have freedom of action within an FEZ to identify and engage air targets in accordance with ROEs in force. If the AAWC wishes to engage a particular target in the FEZ with a long-ange SAM, he is to issue an engagement order assigning the target to an SAM system, ensuring that the fighter controlling agency is so advised. When this order has been issued, a missile arc is automatically established.

Safety Sector

A safety sector is defined by the AAWC and, when required, coordinated with the responsible ACA, in terms of origin, range, center bearing, width, height band, time, and controlling agency. If more than one safety sector is established, they should be numbered and designated. Safety sectors are normally dormant and are activated by the AAWC.

REVIEW QUESTIONS

1. Describe the concept of Defense in Depth.
2. What is the primary objective during the surveillance phase of Air Warfare?
3. What is a picket unit?
4. What is the mission of Combat Air Patrol in Air Warfare?
5. What is PIRAZ?
6. List three considerations that affect AW planning.

CHAPTER 11

Expeditionary Warfare

LEARNING OBJECTIVES

At the end of this chapter the student will be able to:

- Describe how Expeditionary Warfare doctrines contribute to the basic sea control and power projection missions of the naval Service. More specifically:
- Define Expeditionary Warfare. Describe the range of military operations.
- Describe the five elements of Expeditionary Warfare.
- Describe the four reasons for conducting an amphibious operation.
- Describe the four types of amphibious operations.
- Describe the elements of an ATF and the functions of an LF.
- Discuss the five phases of the amphibious assault in their normal sequence.
- Describe the basic structure of the Amphibious Objective Area (AOA).

Additional Required Reading

None.

Introduction

The Expeditionary Warfare tradition in the U.S. Navy is as old as the nation itself. On March 3, 1776, Captain Samuel Nicholas landed his Continental Marine Battalion on New Providence Island in the Bahamas and seized the British fort. After independence, U.S. Navy forces successfully eliminated Barbary Coast extortion in the early 1800s, establishing the precedent for global naval expeditionary operations.

Modern naval Expeditionary Warfare had its origins in exercises conducted during the decades preceding World War II. The exercises identified unique operational and equipment requirements and developed the skills that were perfected in carrier and amphibious operations, which ensured victory in the Pacific Theater.

Modern Expeditionary Warfare

Today, Expeditionary Strike Groups (ESG) carry on the expeditionary tradition of "Maneuver Warfare." They provide the nation a fully ready combat service for forward deployed and able to respond at a moment's notice. Naval Expeditionary Forces were used to restore democracy in Haiti in 1993–4 and to protect U.S. citizens and diplomatic personnel in Monrovia, Liberia, in April 1996. Strikes from the sea punished Iraq for violating U.N. resolutions in 1993 and 1996.

Ready for combat, naval Expeditionary Forces are simultaneously capable of performing large-scale humanitarian operations without reconfiguration. En route home from Operation Desert Storm in 1991, Amphibious Ready Group Three was diverted to Bangladesh, which had been devastated by a tropical cyclone. Navy and Marine Corps Expeditionary Forces protected U.S. citizens in Somalia in 1992 and covered U.N. forces in 1993 and 1995. In late 2001, Marine units deployed from amphibious forces have fought the Taliban and Al–Qaeda in Afghanistan, demonstrating the reach of the Navy and Marine Corps team into landlocked regions as well as coastal ones. Most recently, two Amphibious Task Forces (ATF), embarking two Marine Expeditionary Brigades (MEB), were stationed off the coast of Iraq as part of Operation Iraqi Freedom. The flexibility of these forces allowed the Marines to swiftly stage personnel and equipment in Kuwait and surge into Iraq once the war commenced. All the while, the Marine Air Wings were able to maintain 24/7 operations from the large deck amphibious ships offshore (Sea Basing).

Naval Expeditionary Warfare comprises military operations mounted from the sea, usually on short notice consisting of forward deployed, or rapidly deployable, self-sustaining naval forces tailored to achieve a clearly stated objective. Additionally, Expeditionary Warfare Forces support and perform the full range of military operations on a day-to-day basis.

Naval Expeditionary Forces retain the unique ability to quickly shift operational focus across the spectrum—from combat missions to humanitarian operations—without needing to reconfigure the force.

Peacetime Engagement	Deterrence and Conflict Prevention	Fight and Win
Military-to-Military Contact	Regional Alliances	Clear Objectives— Decisive Force
Assistance to Nations	Crisis Response	Wartime Power Projection
Humanitarian Operations	Confidence-Building Measures	Combined and Joint Warfare
Peacekeeping	Noncombatant Evacuation Operations	Winning the Information War
	Sanctions Enforcement	Counter Weapons of Mass Destruction
	Peacekeeping	Force Generation
		Winning the Peace

Range of Military Operations

There are five principal elements that are vital to mission success in Expeditionary Warfare:

Maritime Dominance—*To* approach land in order to execute an Expeditionary Mission, U.S. Navy and Marine Corps forces must operate in shallow and restricted waters. Control of the sea en route and control of the littoral waters in a large area surrounding the objective is vital to mission success. To get to the objective, our forces will have increasingly fewer forward bases from which to stage. Developments in surveillance will make it increasingly difficult to transit and remain undetected and untargeted. Shallow and restricted waters provide an arena for submarines, torpedoes, and next generation anti-ship cruise missiles. Mines (responsible for the loss or damage of more U.S. Navy ships in recent years than all other weapons combined) will be found throughout the range of coastal waters. Elements of Sea Power 21 are specifically designed to alleviate the threat to our forces as they approach and operate in the littoral regions.

Firepower—*To* be able to detect, identify, categorize, and destroy a target at will is the goal of firepower. It requires sophisticated capabilities in surveillance, fire delivery, and maneuver. Artillery is more self-contained and mobile, enabling artillery and theater ballistic missile batteries to change positions in a few hours, with setup times in minutes. Land-based artillery and rockets often outrange naval guns and can deliver conventional explosives, mines or precision guided munitions. In the next 20 years, laser and energy technologies will transform today's sensor-blinder into a hard-kill weapon. Theater Ballistic Missiles (TBM), already a threat to Naval Expeditionary Forces at the beachhead or in ports, will become accurate enough to target individual ships in amphibious landing areas. Concentration of hostile firepower against Expeditionary Forces will make the environment more challenging. Ongoing developments of ships capable of countering the TBMs are under have been tested and should be operational in the near future.

Maneuver Dominance—*To* gain the initiative critical to mission success, the Expeditionary Commander must have the freedom to maneuver at sea and on land. Natural and man-made obstacles, poor infrastructure, harsh climates, and the combat capabilities of highly mobile armored reaction forces can restrict our ability to maneuver. A turbulent human environment across the range of military operations further limits options. The Marine Corp and Navy are striving to develop ships and systems (Transformation Initiatives) that will afford more flexibility in a constrained environment.

Air Dominance—*Control* of the air over an objective is mandatory, yet potentially unstable nations are importing or building increasingly sophisticated SAMs and integrating them into modern air defense networks. They are also acquiring advance aircraft necessary to patrol air space.

Information Superiority—The commander remains dependent upon the quality and quantity of information that the Command, Control, Communications, Computers, Intelligence, Surveillance, and Reconnaissance assets can deliver. Potential aggressors are using stealth technologies and camouflage, concealment, and deception techniques to hide potential targets. Opposing naval forces blend into the background and clutter; key military facilities and capabilities are buried both physically in the earth, and figuratively within urban areas. Our systems seek them out better than ever before in history, but at the price of demanding a geometric increase in data transfer rates. Communications rely upon satellite links, more efficient use of the electromagnetic spectrum, sophisticated encryption techniques, new transmission devices, and other innovative technologies. All are vulnerable. One of the principal means of conducting Expeditionary Warfare is through the amphibious operation; and in order to be successful, the amphibious operation must employ each of the five elements of Expeditionary Warfare.

Amphibious Operations

An amphibious operation is a military operation launched from the sea by naval and landing forces embarked in ships or craft involving a landing on a hostile or potentially hostile shore. It is directed by the combatant commander, subunified commander, or JTF commander delegated overall responsibility for the operation. An amphibious operation requires extensive air participation and is characterized by closely integrated efforts of forces trained, organized, and equipped for different combat functions.

Amphibious operations are designed and conducted primarily to:

- Prosecute further combat operations.
- Obtain a site for an advanced naval, land, or air base.
- Deny use of an area or facilities to the enemy.
- Fix enemy forces and attention, providing opportunities for other combat operations.

The essential usefulness of an amphibious operation stems from its mobility and flexibility (i.e., the ability to concentrate balanced forces and strike with great strength at a selected point in the hostile defense system). The amphibious operation exploits the element of surprise and capitalizes on enemy weaknesses by projecting and applying combat power at the most advantageous location and time. The threat of an amphibious landing can induce enemies to divert forces, fix defensive positions, divert major resources to coastal defenses (Operation Desert Storm), or disperse forces. Such a threat may result in attempting to defend their coastlines.

The salient requirement of an amphibious assault is the necessity for swift, uninterrupted buildup of sufficient combat power ashore from an initial zero capability to full, coordinated striking power as the attack progresses toward the amphibious task force's (ATF) final objectives.

Types of Amphibious Operation

AMPHIBIOUS ASSAULT—This is the principal type of amphibious operation, which is distinguished from other types of amphibious operations in that it involves establishing a force on a hostile or potentially hostile shore (Inchon, Korea).

AMPHIBIOUS WITHDRAWAL—An amphibious operation involving the extraction of forces by sea in naval ships or craft from a hostile or potentially hostile shore.

AMPHIBIOUS DEMONSTRATION—An amphibious operation conducted to deceive the enemy by a show of force with the expectation of deluding the enemy into a course of action unfavorable to it.

AMPHIBIOUS RAID—An amphibious operation involving swift incursion into or temporary occupation of an objective followed by a planned withdrawal. Raids are conducted for such purposes as:

- Inflicting loss or damage.
- Securing information.
- Creating a diversion.
- Capturing or evacuating individuals and/or material.
- Destroying enemy information gathering systems to support operations security.

The Initiating Directive

The *initiating directive* is the establishing order to the ESG to conduct an amphibious operation. It is used by the combat commander, subunified commander, service component commander, or Joint Task Force (JTF) commander to establish and assign mission parameters such as mission assignments, area of operation, codes, target dates, and special instructions. It may not be a comprehensive document.

Chain of Command

The interrelationship of the Navy and the embarked Marines (MEU Staff) during the planning and execution of an amphibious operation requires the establishment of parallel chains of command and corresponding commanders at all levels of the Expeditionary Strike Group (ESG) organization. Fundamental considerations governing the application of such a system of parallel command include:

Commander, Expeditionary Strike Group, is overall responsible for the planning and execution of all naval ship movement from the staging area to the Amphibious Objective Area (AOA). In addition, the ESG Commander directly oversees all ship to shore movements of personnel and equipment.

Currently, the command structure of the ESG is under study. The West Coast is experimenting with a Flag Officer designated as the ESG Commander and is alternating that post between a Navy and Marine Corps Officer. If this model proves successful, it will be adopted on the East Coast.

Commanding Officer, Marine Expeditionary Unit (MEU), is normally a 0–6 Marine Corps officer, who has Operational Control (OPCON) of the Landing Force (LF). With a traditional ESG (East Coast), the MEU Commander has a co-equal relationship with the ESG Commander on all matters pertaining to the planning and execution of any operations involving the LF. With the West Coast model, the MEU Commander works closely with but is subordinate to the Flag Officer.

The Navy and Marine Corps enjoy a close relationship for planning purposes, and must work together to ensure a good operation. Each brings his/her expertise to the most complex military operation.

Today's Navy/Marine Corps working relationship within an ESG is typified by what is called the "Supported–Supporting" relationship. Simply stated, the Supported Commander is the one who has been designated overall in command of any particular mission. The Supporting Commander will in turn offer all reasonable assistance in order to ensure operational success.

Examples:

If the MEU Commander is tasked to conduct an NEO (Noncombatant Evacuation Operation), the ESG Commander will render all necessary assistance to the MEU. This would obviously include the stationing of the amphibious ships, ship-to-shore movement, etc.

If the ESG Commander is tasked with conducting an MIO (Maritime Interdiction Operation), the MEU Commander will in turn make assets available if so requested. Examples of this might include Marine personnel and or helicopters.

Note: With the current Flag Officer configuration with the West Coast ESGs, there are some significant modifications in the Command and Control relationships. Fundamentally, however, the Supported/Supporting concept is still in play.

Amphibious Task Force Organization

The organization for execution of the amphibious operation reflects interrelationships at every level between the tasks of the LF, corresponding naval forces, special operations forces (SOF), and participating Air Force forces. The interrelationships dictate that special emphasis be given to the parallel chains of command.

The task organization of the ATF must meet the requirements of embarkation, movement to the AOA, protection, landing, and support of the LF. For this reason, the task organization is determined according to the requirements of the anticipated tactical situation. An ESG in most scenarios is sufficient for mission accomplishment; however, a larger force might be required; flexibility is essential.

Amphibious Task Force

The task organization formed for conducting an amphibious operation is the ATF. The ATF always includes Navy forces and an LF, both of which normally have organic aviation assets. Other air and special operations forces (SOF) may be included, as required. Task elements of the ATF are listed below:

- Transport Groups
- Control Group (ship-shore movements)
- TACAIR Control Group
- Naval Surface Fire Support (NSFS) Group
- Carrier Battle Group|Waifare Commanders
- TACAIR Group (shore based)
- Mine Warfare Groups
- SPECWARFARE Groups
- Tactical Deception Groups
- Naval Beach Group
- Construction Battalions

Phases of the Amphibious Assault 117

LANDING FORCES

The LF consists of the command, combat, combat support, and combat service support (CSS) elements assigned to conduct the amphibious assault (air and ground). The LF is specially organized for the following functions:

- Embarkation of troops, equipment, and supplies.
- Debarkation and landing of troops by air and/or surface units.

- Conduct of air and waterborne assault operations.
- Control of Naval Surface Fire Support (NSFS).
- Provision, as appropriate, and control of air support.
- Operation and tactical employment of organic amphibious vehicles and aircraft.
- Discharge of logistics and CSS elements and cargo from assault shipping and establishment of logistical sites and service areas.

Phases of the Amphibious Assault

The amphibious assault follows a well-defined pattern. The general sequence consists of planning, embarkation, rehearsal, movement to the landing area, assault, and accomplishment of the ATF mission. While planning occurs throughout the entire operation, it is normally dominant in the period before embarkation. Successive phases bear the title of the dominant activity taking place within the period covered.

The organization of embarkation needs to provide for maximum flexibility to support alternate plans that may of necessity be adopted. The landing plan and the scheme of maneuver ashore are based on conditions and enemy capabilities existing in the Amphibious Objective Area (AOA) before embarkation of the LF.

Planning

The *planning phase* denotes the period extending from the issuance of the initial directive to embarkation. Although planning does not cease with the termination of this phase, it is useful to distinguish the change that occurs in the relationship between commanders at the time the planning phase terminates and the operational phase begins. Logistics is the science of planning and carrying out the movement and maintenance of forces. In an amphibious operation logistics deals with the design, development, acquisition, storage, movement, distribution, maintenance, evacuation, and disposition of material and personnel. Combat service support (eSS) provides the essential logistics functions and tasks necessary to sustain all elements of operating forces in an area of operation. Planning considerations include assembly and embarkation of personnel and material based on anticipated requirements of LF scheme of maneuver ashore. All considerations that lead to a successful operation should be dealt with in the planning phase, but may be modified by the results of the rehearsal. These considerations include, but are not limited to: the assembly and embarkation of personnel and material based on the anticipated movement ashore *(first off, last on)*, the anticipated strength of the enemy, climate and terrain of the area of operations, the anticipated length of supply lines, communications capabilities, and target dates. Effective logistics and combat service support are absolutely critical to the success of any amphibious operation.

Embarkation

The *embarkation phase* is the period during which the forces, with their equipment and supplies, embark in assigned shipping.

Rehearsal

The *rehearsal phase* is the period during which the prospective operation is rehearsed for the purpose of:

- Testing the adequacy of the plans, the timing of detailed operations, and the combat readiness of participating forces.
- Ensuring all echelons are familiar with plans.
- Testing communications.

Movement

The *movement phase* is the period during which various elements of the ATF move from points of embarkation to the AOA. This move may be via rehearsal, staging, or rendezvous areas. This movement phase is completed when the various elements of the ATF arrive at their assigned positions in the AOA.

Assault

The *assault phase* begins when sufficient elements of the main body of the ATF arrive in assigned positions in the landing area and are capable of beginning the ship-to-shore movement. The assault phase terminates with accomplishment of the ATF mission. The assault phase encompasses:

- Preparation of the landing area by supporting arms.
- Ship-to-shore movement of the LF.
- Air and surface assault landing (assault and initial unloading; tactical) by assault elements of the LF to seize the beachhead and designates ATF and LF objectives.
- Provision of supporting arms and logistics/CSS throughout the assault.
- Landing the remaining elements (general unloading; logistical) for conduct of operations as required for accomplishment of the ATF mission.

PERMA vs. EMPRA

PERMA is the normal doctrinal sequence for conducting an amphibious operation. However, when the organization involved is the ESG and the MEU(SOC), the actual sequence of events that might occur is EMPRA. Simply put, the MEU(SOC) embarks aboard ESG shipping, moves to the area of operations, then receives a mission and conducts the planning, the rehearsal, and the assault. This is due to the forward-deployed nature of the ESGIMEU, which is scheduled months in advance.

Amphibious Objective Area

The Amphibious Objective Area (AOA) is a geographical area, delineated in the initiating directive, for purposes of command and control, within which is located the objectives to be secured by the amphibious task force. This area must be of sufficient size to ensure accomplishment of the amphibious task force's mission and must provide sufficient area for conducting necessary sea, air, and land operations.

Movement of the ATF to the AOA includes departure of ships from loading points in an embarkation area; passage at sea; and approach to, and arrival in, assigned positions in the AOA. The ATF is organized for movement into movement groups, which sail in accordance with the movement plan on prescribed routes. Protective measures are utilized to prevent losses en route. These most often entail combatant escorts and attention to EMCON, tactical evasion and sometimes deception procedures. Movement of the ATF to the AOA may be interrupted by rehearsals, diversion to staging areas for logistics/CSS reasons, or temporary stops at regulating stations or points.

Echelons of the Landing Force

Assault Echelon (AE)

The element of a force that is scheduled for initial assault on the objective area. The AE is embarked on amphibious assault shipping and comprises the tailored units and equipment packages along with maximum amount of supplies that can be loaded to sustain the assault.

Sea Echelon (SE)

A sea echelon is a portion of the assault shipping that withdraws from, or remains out of the transport area during an amphibious landing and operates in designated areas seaward in an on-call or unscheduled status.

Areas Within the AOA

Landing Beach

That portion of the shoreline over which a force approximately the size of a Battalion Landing Team may be landed.

Boat Lane

Lanes in which landing craft and AAVs proceed toward the beach.

NSFS Area

An area in which supporting ships provide cover using missiles and guns.

Helicopter Landing Area (HLA)

A specified ground area for landing assault helicopters to embark or disembark troops and/or cargo. It may contain one or more landing sites.

Transport Area

In amphibious operations, an area assigned to a transport organization for the purpose of debarking troops and equipment. It consists of mineswept lanes, areas, and channels leading to the beaches. The maximum number of ships in the transport area is directly limited by dispersion requirements, availability of forces for MCM operations, and local hydrography and topography.

Sea Echelon Area

An area seaward of the transport area from which assault shipping is phased into the transport area and to which assault shipping withdraws from the transport area.

Operations in the AOA

The CATF controls the assault from the Task Force Command Center aboard the most capable amphibious ship (LCC, LHD, LHA, LPD). He exercises control through the following control centers and group commanders:

SUPPORTING ARMS COORDINATION CENTER (SACC)
- Coordinates pre-assault air strikes, beach clearance operations and gunfire
- Coordinates NSFS, artillery support during assault
- Coordinates helo assault through Helo Coordination Section (HCS)
- Coordinates close air support strikes thru Tactical Air Coordination Center (TACC) and AAW over transport area with the AAWe
- SACC functions shift to Fire Support Control Center (FSCC) ashore under the CLF's control. CLF establishes a TACC, HCS, logistics control centers, and others before assuming full responsibility for future operations ashore.
- Controls movement of landing craft to and from the beach.

BOAT GROUP COMMANDER
TRANSPORT GROUP COMMANDER
- Controls movement and offloading of transports, LSTs, and Assault Ships and assigns Fire Support Areas for NSFS ships.

ASW/AAW/ASUW GROUP COMMANDERS
- Conduct screening operations in the Sea Echelon Area.

Tactical Considerations

Natural Hazards

Surf, weather, and hydrography can often be overcome by careful planning and use of special tactics such as vertical assault, LCACs, underwater demolition, careful timing, etc.

Vulnerability

The landing force is extremely vulnerable in the early hours of the assault as strength ashore must be built up from zero. Careful planning and coordinated execution of the plan is required. The buildup must be rapid and uninterrupted.

Battlespace Dominance

Control of sufficient land, sea, and air space to gain combat superiority in the AOA during the landing is critical. Enemy airfields, shore-based cruise missile sites, and long-range artillery must be neutralized by the preassault strikes and bombardment. The presence of mines and defensive cruise missiles may force the amphibious assault force to keep a longer standoff range and conduct an over-the-horizon assault.

REVIEW QUESTIONS

1. What is Maritime Dominance and how can it be achieved?
2. List two reasons for conducting an amphibious operation.
3. What is a raid and how is it different from an assault?
4. What is the purpose of Initiating Directive as it relates to planning an amphibious operation?
5. Describe the command relationship between the CATF and CLF.
6. List four of the elements that make up an ATF.
7. What are the phases of the amphibious assault in their normal sequence?
8. Describe the areas/zones of the Amphibious Objective Area.
9. Why is Battlespace Dominance an important tactical consideration in Amphibious Operations?

SUGGESTED FURTHER READING

Joint Pub 3–02: *Joint Doctrine for Amphibious Operations (1992)* Government Printing Office, Washington D.C.

NAVEDTRA 10776–A: *Surface Ship Operations* Naval Education and Training Command Government Printing Office, Washington D.C.

Wadle, S. Midn, USN *Operation Roman Candle* [NS 310 Project], U. S. Naval Academy Annapolis MD Worldwide Challenges to Naval Expeditionary Warfare, Office of Naval Intelligence, March 1997.

CHAPTER 12

Marine Corps Warfighting

LEARNING OBJECTIVES

At the end of this chapter the student will be able to:

- Decision Process
- Philosophy of Command
- Orders
- Commander's Intent
- Describe the concept and the different types of MAGTF
- Demonstrate how to estimate a situation using METT-T
- Utilize the concepts of maneuver warfare in a Tactical Decision Game (TDG)

ADDITIONAL READING

MCDP 1: *Warfighting*, Department of the Navy, Headquarters Marine Corps, Washington, D.C., 1989.

MCDP 1-3: *Tactics*, Department of the Navy, Headquarters Marine Corps, Washington, D.C., 1997.

Introduction

To understand the Marine Corps' philosophy of warfighting, we first need an appreciation for the nature of war itself. War is a state of hostilities that exists between or among nations, characterized by the use of military force. The aim in war is to impose our will on the enemy. We must either eliminate his physical ability to resist, or short of this, (we must deploy his will to resist.) The sole justification for the U.S. Marine Corps is to secure or protect national policy objectives by military force when peaceful means alone cannot.

Warfighting

The challenge is to identify and adopt a concept of warfighting consistent with our understanding of the nature and theory of war and the realities of the modern battlefield. This requires a concept of warfighting that will function in an uncertain, chaotic, and fluid environment—in fact, one that will exploit

Two landing crafts, air-cushioned, land on a beach in Sanmaesan, Thailand during amphibious landing training for exercise Cobra Gold 2009. —Lance Cpl. Daniel A. Flynn

these conditions to advantage. It requires a concept that generates and exploits superior tempo and velocity. It requires a concept that is consistently effective across the full spectrum of conflict, because we cannot attempt to change our basic doctrine from situation to situation and expect to be proficient. It requires a concept which recognizes and exploits the fleeting opportunities which naturally occur in war. It requires a concept which takes into account the moral as well as the physical forces of war. It requires a concept with which we can succeed against a numerically superior foe, because we can no longer presume a numerical advantage. And, especially in expeditionary situations in which public support for military action may be tepid and short lived, it requires a concept with which we can win quickly against a larger foe on his home soil, with minimal casualties and external support.

Maneuver Warfare

The Marine Corps concept for winning under these conditions is a warfighting doctrine based on rapid, flexible, and opportunistic maneuver. In order to fully appreciate what we mean by maneuver we need to clarify the term. The traditional understanding of maneuver is a spatial one; that is, we maneuver in space to gain a positional advantage. However, in order to maximize the usefulness of maneuver, we must consider maneuver *in time* as well; that is, we generate a faster operational tempo than the enemy to gain a temporal advantage. It is through maneuver in dimensions that an inferior force can achieve decisive superiority at the necessary time and place.

From this point of view we see that the aim in maneuver warfare is to render the enemy incapable of resisting by shattering his moral and physical cohesion—his ability to fight as an effective, coordinated whole—rather than to destroy him physically through incremental attrition, which is generally more costly and time-consuming. Ideally, the components of his physical strength that remain are irrelevant because we have paralyzed his ability to use them effectively. Even if an outmaneuvered enemy continues to fight as individuals or small units, we can destroy the remnants with relative ease because we have eliminated his ability to fight effectively as a force.

This is not to imply that firepower is unimportant. On the contrary, the suppressive effects of firepower are essential to our ability to maneuver. Nor do we mean to imply that we will pass up the opportunity to physically destroy the enemy. We will concentrate fires and forces at decisive points to destroy enemy elements when the opportunity presents itself and when it fits our larger purposes. But the aim is not an unfocused application of firepower for the purpose of incrementally reducing the enemy's physical strength. Rather, it is the *selective* application of firepower in support of maneuver to contribute to the enemy's shock and moral disruption. The greatest value of firepower is not physical destruction—the cumulative effects of which are felt only slowly—but the moral dislocation it causes.

If the aim of maneuver warfare is to shatter the enemy's cohesion, the immediate object toward that end is to create a situation in which he cannot function. By our actions, we seek to pose menacing dilemmas in which events happen unexpectedly and faster than the enemy can keep up with them. The enemy must be made to see his situation not only as deteriorating, but deteriorating at an ever-increasing rate. The ultimate goal is panic and paralysis, an enemy who has lost the ability to resist.

Inherent in maneuver warfare is the need for speed to seize the initiative, dictate the terms of combat, and keep the enemy off balance, thereby increasing his friction. Through the use of greater tempo and velocity, we seek to establish a pace that the enemy cannot maintain so that with each action his reactions are increasingly late—until eventually he is overcome by events.

Also inherent is the need for violence, not so much as a source of physical attrition but as a source of moral dislocation. Toward this end, we concentrate strength against *critical* enemy vulnerabilities, striking quickly and boldly where, when, and how it will cause the greatest damage to our enemy's ability to fight. Once gained or found, any advantage must be pressed relentlessly and unhesitatingly. We must be ruthlessly opportunistic, actively seeking out signs of weakness, against which we will direct all available combat power. And when the *decisive* opportunity arrives, we must exploit it fully and aggressively, committing every ounce of combat power we can muster and pushing ourselves to the limits of exhaustion.

The final weapon in our arsenal is surprise, the combat value of which we have already recognized. By studying our enemy we will attempt to appreciate his perceptions. Through deception we will try to shape his expectations. Then we will dislocate them by striking at an unexpected time and place. In order to appear unpredictable, we must avoid set rules and patterns, which inhibit imagination and initiative. In order to appear ambiguous and threatening, we should operate on axes that offer several courses of action, keeping the enemy unclear as to which we will choose.

Philosophy of Command

It is essential that our philosophy of command support the way we fight. First and foremost, *in order to generate the tempo of operations we desire and to best cope with the uncertainty, disorder, and fluidity of combat, command must be decentralized.* That is, subordinate commanders must make decisions on their own initiative, based on their understanding of their senior's intent, rather than passing information up the chain of command and waiting for the decision to be passed down. Further, a competent subordinate commander who is at the point of decision will naturally have a better appreciation for the true situation than a senior some distance removed. Individual initiative and responsibility are of paramount importance. The principal means by which we implement decentralized control is through the use of mission tactics, which we will discuss in detail later.

Second, since we have concluded that war is a human enterprise and no amount of technology can reduce the human dimension, our philosophy of command must be based on human characteristics rather than on equipment or procedures. Communications equipment and command staff procedures can enhance our ability to command, but they must not be used to replace the human element of command. Our philosophy must not only accommodate, but must exploit human traits such as boldness, initiative, personality, strength of will, and imagination.

Our philosophy of command must also exploit the human ability to communicate *implicitly*. We believe that *implicit communication, to* communicate through *mutual understanding,* using a minimum of key, well-understood phrases or even anticipating each other's thoughts, is a faster, more effective way to communicate than through the use of detailed, explicit instructions. We develop this ability through familiarity and trust, which are based on a shared philosophy and shared experience.

This concept has several practical implications. First, we should establish long-term working relationships to develop the necessary familiarity and trust. Second, key people, "actuals," should talk directly to one another when possible, rather than through communicators or messengers. Third, we should communicate orally, when possible, because *how* we talk; our inflections and tone of voice are also important forms of communication. And fourth, we should communicate in person whenever possible, because we communicate also through our gestures and bearing.

A commander should command from well forward. This allows him to see and sense firsthand the ebb and flow of combat, to gain an intuitive appreciation for the situation which he cannot obtain from reports. It allows him to exert his personal influence at decisive points during the action. It also allows him to locate himself closer to the events that will influence the situation so that he can observe them directly and circumvent the delays and inaccuracies that result from passing information up the chain of command. Finally, we recognize the importance of personal leadership. Only by his physical presence, by demonstrating the willingness to share danger and privation, can the commander fully gain the trust and confidence of his subordinates.

As part of our philosophy of command we must recognize that war is inherently disorderly, uncertain, dynamic, and dominated by friction. Moreover, maneuver warfare, with its emphasis on speed and initiative, is by nature, a particularly disorderly style of war. The conditions ripe for exploitation are normally also very disorderly. For commanders to try to gain certainty as a basis for actions, maintain positive control of events at all times, or shape events to fit their plans is to deny the very nature of war. We must therefore be prepared to cope, even better, to *thrive,* in an environment of chaos, uncertainty, constant change, and friction. If we can come to terms with those conditions and thereby limit their debilitating effects, we can use them as a weapon against a foe who does not cope as well.

In practical terms this means that we must not strive for certainty before we act for in doing so we will surrender the initiative and pass up opportunities. We must not try to maintain positive control over subordinates since this will necessarily slow our tempo and inhibit initiative. We must not attempt to impose precise order to the events of combat since this leads to a formulistic approach to war. And we must be prepared to adapt to changing circumstances and exploit opportunities as they arise, rather than adhering insistently to predetermined plans.

There are several points to remember about our command philosophy. First, while it is based on our warfighting style, this does not mean it applies only during war. We must put it into practice during the preparation for war as well. We cannot rightly expect our subordinates to exercise boldness and initiative in the field when they are accustomed to being over supervised in the rear. Whether the mission is training, procuring equipment, administration, or police call, this philosophy should apply.

Next, our philosophy requires competent leadership at all levels. A centralized system theoretically needs only one competent person, the senior commander, since his is the sole authority. But a decentralized system requires leaders at all levels to demonstrate sound and timely judgment. As a result, initiative becomes an essential condition of competence among commanders.

Our philosophy requires familiarity among comrades because only through a shared understanding can we develop the implicit communication necessary for unity of effort. Finally, and, most important, ***our philosophy demands confidence among seniors and subordinates.***

Decision Making

Decision making is essential to the conduct of war since all actions are the result of decision, or nondecision. If we fail to make a decision out of lack of will, we have willingly surrendered the initiative to our foe. If we consciously postpone taking action for some reason, that is a decision. Thus, as a basis for action, any decision is generally better than no decision.

Since war is a conflict between opposing wills, we cannot make decisions in a vacuum. We must make our decisions in light of the enemy's anticipated reactions and counteractions, recognizing that while we are trying to impose our will on our enemy, he is trying to do the same to us.

Whoever can make and implement his decisions consistently faster gains a tremendous, often decisive, advantage. Decision making thus becomes a time-competitive process, and timeliness of decisions becomes essential to generating tempo. Timely decisions demand rapid thinking, with consideration limited to essential factors. We should spare no effort to accelerate our decision-making ability.

A military decision is not merely a mathematical computation. Decision making requires both the intuitive skill to recognize and analyze the essence of a given problem and the creative ability to devise a practical solution. This ability is the product of experience, education, intelligence, boldness, perception, and character.

We should base our decisions on *awareness* rather than on mechanical *habit*. That is, we act on a keen appreciation for the essential factors that make each situation unique instead of form conditioned response.

We must have the moral courage to make tough decisions in the face of uncertainty—and accept full responsibility for those decisions—when the natural inclination would be to postpone the decision pending more complete information. To delay action in an emergency because of incomplete information shows a lack of moral courage. We do not want to make rash decisions, but we must not squander opportunities while trying to gain more information.

We must have the moral courage to make bold decisions and accept the necessary degree of risk when the natural inclination is to choose a less ambitious tack, for "in audacity and obstinacy will be found safety."

Finally, since all decisions must be made in the face of uncertainty and since every situation is unique, there is no perfect solution to any battlefield problem. Therefore, we should not agonize over one. The essence of the problem is to select a promising course of action with an acceptable degree of risk, and to do it more quickly than our foe. In this respect, "a good plan violently executed *now* is better than a perfect plan executed next week."

Mission Tactics

Having described the object and means of maneuver warfare and its philosophy of command, we will next discuss how we put maneuver warfare into practice. First is through the use of mission tactics. Mission tactics are just as the name implies: *the tactic of assigning a subordinate mission without specifying how the mission must be accomplished.* We leave the manner of accomplishing the mission to the subordinate, thereby allowing him the freedom and establishing the duty to take whatever steps he deems necessary based on the situation. The senior prescribes the method of execution only to the degree that is essential for coordination. It is this freedom for initiative that permits the high tempo of operations that we desire. Uninhibited by restrictions from above, the subordinate can adapt his actions to the changing situation. He informs his commander what he has done, but he does not wait for permission.

It is obvious that we cannot allow decentralized initiative without some means of providing unity, or focus, to the various efforts. To do so would be to dissipate our strength. We seek unity, not through imposed control, but through *harmonious* initiative and lateral coordination.

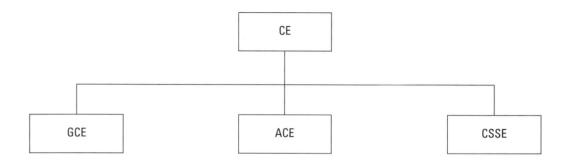

Commander's Intent

We achieve this harmonious initiative in large part through the use of the commander's intent. There are two parts to a mission: the task to be accomplished and the reason, or intent. The task describes the desired result of the action. Of the two, the intent is predominant. While a situation may change, making the task obsolete, the intent is more permanent and continues to guide our actions. Understanding our commander's intent allows us to exercise initiative in harmony with the commander's desires.

In order to maintain our focus on the enemy, we should try to express intent in terms of the enemy. The intent should answer the question: *What do I want to do to the enemy?* This may not be possible in all cases, but it is true in the vast majority. The intent should convey the commander's *vision.* It is not satisfactory for the intent to be "to defeat the enemy." To win is always our ultimate goal, so an intent like this conveys nothing.

From this discussion, it is obvious that a clear explanation and understanding of intent is absolutely essential to unity of effort. It should be a part of any mission. The burden of understanding falls on senior and subordinate alike. The senior must make perfectly clear the result he expects, but in such a way that does not inhibit initiative. Subordinates must have a clear understanding of what their commander is thinking. Further, they should understand the intent of the commander two levels up. In other words, a platoon commander should know the intent of his battalion commander, a battalion commander the intent of his division commander.

Marine Corps Organization—The MAGTF Concept

MARINE AIR-GROUND TASK FORCE (MAGTF)

The MAGTF is the force that the Marine Corps employs to conduct maneuver warfare in a combined arms operation. The nature of the MAGTF—cohesion, unity of effort, flexibility, and self-sustainment makes it equal to the requirements of combined arms warfare. The MAGTF contains four elements that can be tailored to a combined arms operation: a command element, a ground combat element, an aviation combat element, and a combat service support element. The MAGTF draws forces from ground, aviation, and combat service support organizations of the Fleet Marine Force (FMF) to meet this requirement.

SIX SPECIAL CORE COMPETENCIES

MAGTF operations are built on a foundation of six special core competencies:

1. Expeditionary readiness
2. Expeditionary operations
3. Combined-arms
4. Forcible entry from the sea

5. Sea-based operations
6. Reserve integration:
   a. Other units (MAGTFs)
   b. Joint or coalition forces.

TYPES OF MAGTF

MAGTFs range in size from the smallest (which can number from fewer than 100 to 3,000 Marines) to the largest (which can number from 40,000 to 100,000 Marines). There are four basic sizes/types of MAGTF:

1. Marine Expeditionary Force (MEF)
2. Marine Expeditionary Brigade (MEB)
3. Marine Expeditionary Unit (MEU)
4. Special Purpose MAGTF (SPMAGTF)

MAGTF COMPOSITION

Regardless of the size of the MAGTF, all have the same basic structure. There are four elements of an MAGTF: the command element (CE), the ground combat element (GCE), the aviation combat element (ACE), and the combat service support element (CSSE).

1. Command Element (CE). The command element is task-organized to provide command and control capabilities (including intelligence and communications) necessary for effective planning, direction, and execution of all operations.

   a. Composition of the CE
      (a) MAGTF Commander
      (b) Deputy Commander
      (c) General Staff
      (d) Special Staff

   b. Functions of the CE. Several key aspects of the CE activities are different from those of its major subordinate commands.
      (a) Drive operations
      (b) Requesting and integrating joint capabilities
      (c) Collecting intelligence
      (d) Deep, close, and rear operations
      (e) Deception and psychological operations
      (f) NBC weapon systems
      (g). Command, control, communications, and intelligence
      (h). MAGTF concept of operations
      (i). Task organizing the MAGTF forces

2. Ground Combat Element (GCE). The GCE is task-organized to conduct ground operations in support of the MAGTF mission. It is normally formed around an infantry organization reinforced with requisite artillery, reconnaissance, armor, and engineer forces and can vary in size and composition from a rifle platoon to one or more Marine divisions. It has some organic combat service support capability.

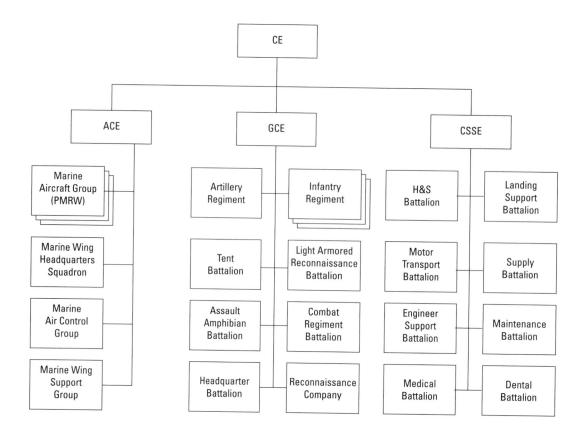

3. Aviation Combat Element (ACE). The ACE is task-organized to support the MAGTF mission by performing some or all of the six functions of Marine aviation. It is normally built around an aviation organization that is augmented with appropriate air command and control, combat, combat support, and CSS units. The ACE can operate effectively from ships, expeditionary airfields, or austere forward operating sites and can readily and routinely transition between sea bases and expeditionary airfields without loss of capability. The ACE can vary in size and composition from an aviation detachment with specific capabilities to one or more Marine Air Wings (MAW).

4. Combat Service Support Element (CSSE). The CSSE is task-organized to provide the full range of CSS functions and capabilities needed to support the continued readiness and sustainability of the MAGTF as a whole. It is formed around a CSS headquarters and may vary in size and composition from a support detachment to one or more Marine Force Service Support Group (FSSG)

MARINE EXPEDITIONARY FORCE (MEF)

An MEF is the *largest and most capable MAGTF.* Because the MEF can deploy with a formidable fighting force that can sustain itself, it is the Marine Corps' *"Force of Choice."* It is normally composed of one or more Marine divisions, Marine air wings, and Force service support groups. A lieutenant general normally commands an MEF. *It comes with 60 days of sustainment* and the *CE is capable acting as a joint/ combined headquarters.*

1. Permanent MEF Headquarters:

(a) I MEF—Camp Pendleton, California;

(b) II MEF—Camp Lejeune, North Carolina;

(c) III MEF—Okinawa, Japan;

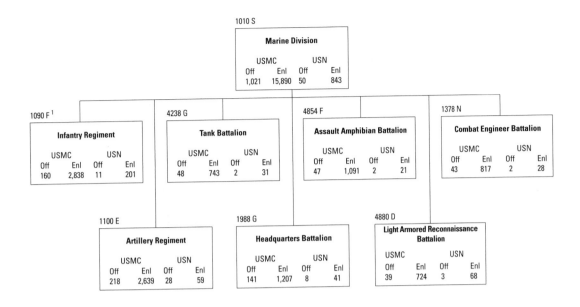

2. Notional MEF. An MEF's typical composition provides for the following:

(a) Marine Division (MarDiv)

The MarDiv is the largest permanent organization of ground combat power in the Fleet Marine Force.

A MarDiv may be employed as the GCE of a large landing force or provide Regimental (RLT) and/or Battalion Landing Teams (BLT) for employment with smaller landing forces. Major subordinate elements of the MarDiv are:

1. Infantry Regiment (X3)
2. Artillery Regiment
3. Tank Battalion
4. Light Armor Reconnaissance Battalion
5. Assault Amphibian Battalion
6. Combat Engineer Battalion
7. Headquarters Battalion

(b) Marine Aircraft Wing (MAW)

The MAW is the largest organization of aviation combat power in the FMF. There are three active duty MAWs and one reserve.

An MAW, which is commanded by a Major General, may be employed as the ACE of a large landing force or provide composite Marine Aircraft Groups and/or squadrons to be employed with smaller landing forces. Major subordinate elements of the MAW are:

1. Marine Air Groups (MAG)s:
   a. Fixed-wing squadrons (VMFA, VMA, VMGR).
   b. Rotary-wing squadrons (HMH, HMM, HML, HML/A).
   c. Unmanned Aerial Vehicles (VMU).

2. Marine Wing Support Group (MWSG). Provides all essential ground support requirements to aid designated fixed-or rotary-wing components.

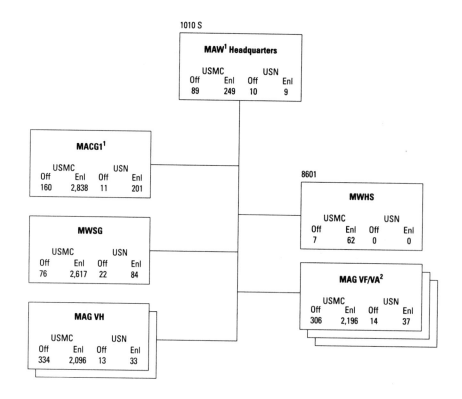

3. Marine Air Control Group (MACG). Its mission is to provide, operate, and maintain the Marine Air Command and Control System (MACCS). It coordinates all aspects of air command and control and air defense within the MAW.
   a. Marine Tactical Air Command Squadron/Tactical Air Command Center (TACC)
   b. Marine Air Control Squadron/Tactical Air Operations Center (TAOC)
   c. Marine Air Support Squadron/Direct Air Support Center (DASC)
   d. Low-Altitude Air Defense Battalion/Battery (LAAD Bn/BtryStinger/60 Avengers)
   e. Marine Wing Communications Squadron (MWCS)

4. Marine Wing Headquarters Squadron (MWHS).

(c) Force Service Support Group (FSSG)

FSSG is the largest composite grouping of combat service support units in the FMF. There are three active-duty FSSGs and one reserve FSSG. Each FSSG, which is normally commanded by a Brigadier General, may be employed as the CSSE of a large landing force or provide a task-organized CSSE for employment with smaller landing forces. Major subordinate units of the FSSG are:

1. Headquarters and Service Battalion
2. Engineer Support Battalion
3. Medical Battalion
4. Dental Battalion
5. Maintenance Battalion
6. Transportation Support Battalions
7. Supply Battalion

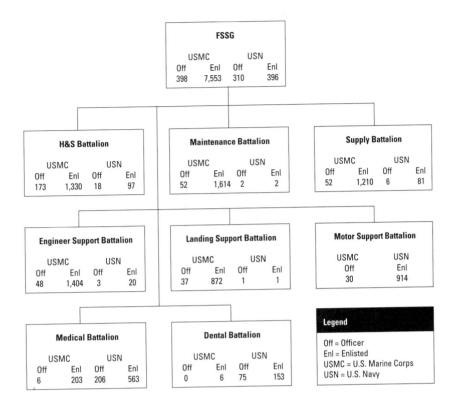

(d) Marine Expeditionary Brigade (MEB)

The Marine Expeditionary Brigade (MEB) is the mid-sized MAGTF and is normally commanded by a brigadier general The MEB bridges the gap between the MEV, at the tip of the spear, and the MEF, our principal warfighter. With 30 days of sufficient supplies for sustained operations, the MEB is capable of conducting amphibious assault operations and maritime prepositioning force (MPF) operations. During potential crisis situations, an MEB may be forward deployed afloat for an extended period in order to provide an immediate combat response.

An MEB can operate independently or serve as the advance echelon of an MEF. The MEB CE is embedded in the MEF CE and identified by line number for training and rapid deployment. The MEB can provide supported CINCs with a credible warfighting capability that is rapidly deployable and possesses the capability to impact all elements of the battlespace. If required, an MEB CE is capable of assuming the role of JTF Headquarters for small operations with additional MEF CE augmentation. As an expeditionary force, it is capable of rapid employment and employment via amphibious shipping, strategic air/sealift, geographical or maritime prepositioning force assets, or any combination thereof. There are three standing MEB command elements: 1st Marine Expeditionary Brigade, assigned within I Marine Expeditionary Force, and located at Camp Pendleton, CA; 2d Marine Expeditionary Brigade, assigned within II Marine Expeditionary Force, and located at Camp Lejeune, NC; and 3d Marine Expeditionary Brigade, assigned within III Marine Expeditionary Force, and located in Okinawa, Japan. 1st and 2d MEB CEs were activated in November 1999. 3d MEB CE was activated in January 2000.

The composition of an MEB varies according to the mission, forces assigned, and the area of operations. An MEB is typically organized with the following elements:

MEB Command Element. The MEB command element will provide command and control for the elements of the MEB. When missions are assigned, the notional MEB command element is tailored with required support to accomplish the mission. Detachments are assigned, as necessary, to support subordinate elements. The MEB CE is fully capable of executing all of the staff functions of a MAGTF (administration and personnel, intelligence, operations and training, logistics, plans, communications and information systems, PAD, SJA, Comptroller, and COMSEC).

Ground Combat Element (GCE). The ground combat element (GCE) is normally formed around a reinforced infantry regiment. The GCE can be composed of from two to five battalion sized maneuver elements (infantry, tanks, LAR) with a regimental headquarters, plus artillery, Assault Amphibian BN, reconnaissance, TOWs, and engineers.

Aviation Combat Element (ACE). The aviation combat element (ACE) is a composite Marine aircraft group (MAG) task-organized for the assigned mission. It usually includes both helicopters and fixed-wing aircraft, and elements from the Marine wing support group and the Marine air control group. The MAG has more varied aviation capabilities than those of the aviation element of a MEU. The most significant difference is the ability to command and control aviation with the Marine Air Command and Control System (MACCS). The MAG is the smallest aviation unit designed for independent operations with no outside assistance except access to a source of supply. Each MAG is task-organized for the assigned mission and facilities from which it will operate. The ACE headquarters will be an organization built upon an augmented MAG.

Combat Service Support Element (CSSE). The brigade service support group (BSSG) is task-organized to provide CSS beyond the capability of the supported air and ground elements. It is structured from personnel and equipment of the force service support group (FSSG). The BSSG provides the nucleus of the landing force support party (LFSP) and, with appropriate attachments from the GCE and ACE, has responsibility for the landing force support function when the landing force shore party group is activated.

The MEB is deployed via a continuous flow of task-organized forces building on MAGTFs. As an expeditionary force, it is capable of rapid deployment and employment via amphibious shipping, strategic air/sealift, marriage with geographical or maritime prepositioning force assets, or any combination thereof. The MEB deploys with sufficient supplies to sustain operations for 30 days. The MEB may be comprised of elements from MPF, ACF, or the ATF. Early command and control forward is critical; therefore an MEB will be deployed with enabling communications into theater as quickly as possible. The MEB provides operational agility to the MEB Commander and supports all warfighting functions: maneuver, intelligence, logistics, force protection, fires, and command headquarters or provided from other MAW assets.

Marine Expeditionary Unit (Special Operations Capable) MEU(SOC). In 1983, the Secretary of Defense directed each military Service and defense agency to review their existing special operations capabilities and develop a plan for achieving the level of special operations capability required to combat both current and future low intensity conflicts and terrorist threats. In response, the Marine Corps instituted an aggressive SOC training program to optimize the inherent capability of MEUs to conduct selected maritime special operations.

Progressive improvement in individual and unit skills attained through enhanced training and the addition of specialized equipment allow an MEU to execute a full range of conventional and selected maritime special operations. This is accomplished by means of dedicated and intensive pre-deployment

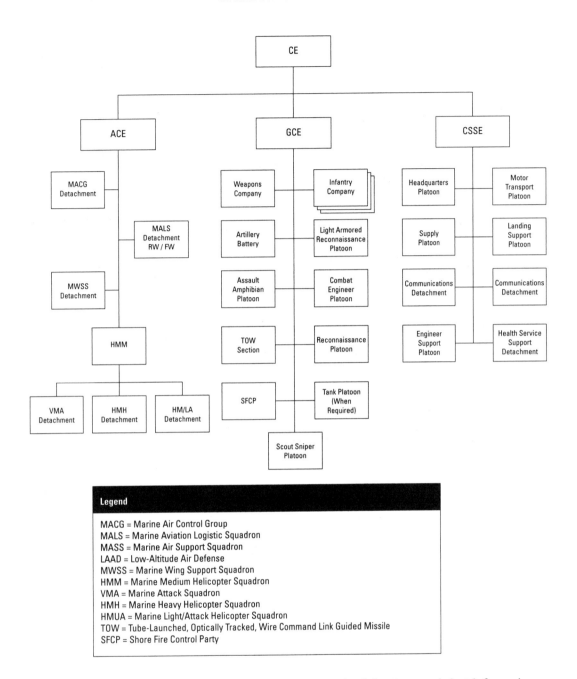

Legend
MACG = Marine Air Control Group
MALS = Marine Aviation Logistic Squadron
MASS = Marine Air Support Squadron
LAAD = Low-Altitude Air Defense
MWSS = Marine Wing Support Squadron
HMM = Marine Medium Helicopter Squadron
VMA = Marine Attack Squadron
HMH = Marine Heavy Helicopter Squadron
HMUA = Marine Light/Attack Helicopter Squadron
TOW = Tube-Launched, Optically Tracked, Wire Command Link Guided Missile
SFCP = Shore Fire Control Party

training program of about 26 weeks that emphasizes personnel stabilization coupled with focused, standardized, and integrated Amphibious Ready Group (ARG)/MEU training. MEUs that have undergone this enhanced training program have been provided special equipment, and have successfully completed a thorough evaluation/certification under the cognizance of the Force Commander, shall be designated as SOC. The primary for all MEUs shall be certification and designation as SOC prior to deployment.

The *primary objective* of the MEU(SOC) is to provide the National Command Authorities and geographic combatant commanders with an effective means of dealing with the uncertainties of future threats, by providing forward-deployed units which offer unique opportunities for a variety of quick reaction, sea-based, crisis response options, in either a conventional amphibious role, or in the execution of selected maritime special operations. "From the Sea" articulates the vision of MAGTFs participating in naval expeditionary forces of combined arms, which are task-organized, equipped, and trained to conduct forward presence and crisis response missions while operating in littoral areas of the world.

Organization of the Marine Expeditionary Unit (Special Operations Capable). The forward-deployed MEU (SOC) is uniquely organized and equipped to provide the naval or joint force commander with rapidly deployable, sea-based capability with 15 days of sustainment optimized for forward presence and crisis response missions. The MEU (SOC) may also serve as an enabling force for follow-on MAGTFs (or possibly joint/combined forces) in the event the situation or mission requires additional capabilities and resources.

The MEU is comprised of a command element; a reinforced infantry battalion as the GCE; a composite helicopter squadron as the ACE; and a CSSE designated the MEU Service Support Group (MSSG). Currently there are seven permanent MEUs: 11th, 13th, and 15th on the West Coast at Camp Pendleton, CA.; 22d, 24th, and 26th on the East Coast at Camp Lejeune, NC and the 31st MEU in Okinawa, Japan. All MEUs have their own identical table of organization, table of equipment, and a separate monitor command code. Most importantly, there are always two deployed MEUs, two deploying MEUs, and two MEUs doing the 26-week "work-up" to deployment. 31st MEU in Okinawa has recently begun participation in the regular 6-month deployment rotation.

Command Element. The CE of the MEU (SOC) is a permanently established organization augmented to provide the command and control (C2) functions and the command, control, communications, computers and intelligence systems (C4I) necessary for effective planning and execution of all operations. In addition to permanently assigned Marines, the *MEU* CE is augmented with detachments from the MEF Headquarters Group (MHG) for deep reconnaissance, fire support, intelligence, electronic warfare, and communications.

MEU STAFF consists of Headquarters Section, Administration Section *(S-1),* including Staff Judge Advocate, Operations Section *(S-3),* Intelligence Section *(S-2),* Logistics Section *(S-4),* and Communications Section *(S-6).*

Maritime Special Purpose Force (MSPF). The MSPF is a unique task organization drawn from the *MEU* major subordinate elements. The MSPF is not designed to duplicate existing capabilities of Special Operation Forces, but is intended to focus on operations in a maritime environment. The MSPF provides the enhanced operational capability to complement or enable conventional operations or to execute special maritime operations. The MSPF cannot operate independently of its parent *MEU.* It relies on the *MEU* for logistics, intelligence, communications, transportation, and supporting fires. Accordingly, command of the MSPF must remain under the control of the *MEU* commander. The MSPF is organized and trained to be rapidly tailored to meet a specific mission. It is notionally comprised of:

1. Command Element. Commander, Comm. Det, Marine Liaison Group Det, Medical Section, Interrogator/Translator Team (ITT) Det and Counterintelligence (CI) Det.
2. Covering Element. Structured around rifle platoon provided by the Battalion Landing Team and may be augmented by the Naval Special Warfare Task Unit (NSWTV). The covering element will act as a reinforcing unit, a support unit, a diversionary unit, or an extraction unit.
3. Strike Element. Focus of effort of the MSPF and is organized to perform assault, explosive breaching, internal security, and sniper functions. FORECON Det, Security Teams, EOD Det, Combat Photo Team, and possibly an NSWTV.
4. Aviation Support Element. Provided by the ACE. Specific structure will vary, but will have the capabilities of precise night flying and navigation, plus various insertion/extraction means and forward area refueling point (FARP) operations.

5. Reconnaissance and Surveillance Element. Normally composed of assets from the Battalion Landing Team (BLT) STA platoon (sniper support) coupled with elements of the RADBN Det, COMM Det, MLG Det, and CI/ITT assets from the *MEU* CE.

Ground Combat Element (GCE) The GCE is normally structured around a reinforced infantry battalion that forms a BATTALION LANDING TEAM (BLT). Specific reinforcements will vary, but generally include artillery, reconnaissance, light armor (maybe tanks), anti-armor, amphibious assault vehicles, and combat engineer attachments. The battalion consists of an H&S company, three letter companies, and a weapons company. There are two important things to note about the BLT, first, unlike a standard infantry battalion, the BLT (when formed) comes to full strength in personnel and equipment (T/O & *TIE*). Secondly, the companies within the BLT have become specialized. One company in the BLT specializes in Mechanized operations, one company specializes in Helo Operations, while the third company specializes in small boat operations.

Aviation Combat Element (ACE) The ACE is a reinforced helicopter squadron that includes AV-8B Harrier attack aircraft, and two CONUS based KC-130 aircraft. The ACE is task-organized to provide assault support, fixed-wing and rotary wing close air support, airborne command and control, and low-level, close-in air defense. The ACE is structured as follows:

1. HMM SQUADRON: Marine Medium Helicopter Squadron configured with twelve CH-6E helicopters: provides medium-lift assault support.
2. HMH DET: Marine Heavy Helicopter Squadron detachment configured with CH-53E helicopters: provides extended range, heavy-lift assault.
3. HMLlA DET: Marine Light Attack Squadron detachment configured with four AH-IW attack helicopters, and three UN-IN utility helicopters: provides close air support, airborne command and control, and escort.
4. VMA DET: Marine Attack Squadron detachment configured with six AV-8B Harrier aircraft: provides organic close air support. When appropriate shipping (i.e., LHA, LHD) is not available, the detachment trains with the MEU throughout pre-deployment training, and then is placed on CONUS standby and prepared to deploy within 96 hours.
5. VMGR DET: Marine Aerial Refueler Transport Squadron detachment configured with two KC-I30 aircraft: provides refueling services for embarked helicopters and AV-8B aircraft, and performs other tasks (i.e., parachute operations, flare drops, cargo transportation, etc.) as required. The detachment trains with the MEU throughout pre-deployment training, and then is placed on CONUS standby and prepared to deploy within 96 hours.
6. MARINE AIR CONTROL GROUP DET
   a. LOW-ALTITUDE AIR DEFENSE (LAAD) BATTALION DET: Provides low level, close-in air defense utilizing MANPAD and the Avenger Stinger Missile Systems.
   b. MARINE AIR SUPPORT SQUADRON DET.: Provides a limited Direct Air Support Center (DASC) capability for enhanced integration of air support into the MEU(SOC) scheme of maneuver.
7. MARINE WING SUPPORT SQUADRON DET: Provides aviation bulk fuel and limited food service support.
8. MARINE AVIATION LOGISTICS SQUADRON DET: Provides intermediate maintenance and aviation supply support.

Combat Service Support Element (CSSE). The CSSE is an MEU Service Support Group (MSSG) which provides the full range of combat service support including supply, maintenance, transportation, deliberate engineering, medical and dental, automated information processing, utilities, landing support (port/airfield support operations), disbursing, legal, and postal services and 15 days of sustainability (Class I, II, III (B), IV, V, VIII, IX) necessary to support MEU (SOC) assigned missions.

Missions of an MEU (SOC): The MEU (SOC) is a self-sustained, amphibious, combined arms air-ground task force capable of conventional and selected maritime special operations of limited duration in support of a Combatant commander. The following is the mission statement from MCO 3120.9A:

To provide the geographic combatant commander a forward-deployed, rapid crisis response capability by conducting conventional amphibious and selected maritime special operations under the following conditions: at night; under adverse weather conditions; from over the horizon; under emissions control; from the sea, by surface and/or by air; commence execution within 6 hours of receipt of the warning order. To act as an enabling force for a follow-on MAGTF or joint and/or combined forces in support of various contingency requirements.

The inherent capabilities of a forward-deployed MEU (SOC) are divided into four broad categories:

1. Amphibious Operations
   (a) Amphibious Assault
   (b) Amphibious Raid
   (c) Amphibious Demonstration
   (d) Amphibious Withdrawal

2. Direct Action Operations
   (a) In-Extremis Hostage Recovery (IHR)
   (b) Seizure/Recovery of Offshore Energy Facilities.
   (c) Visit, Board, Search and Seizure Operations (VBSS)
   (d) Specialized Demolition Operations
   (e) Tactical Recovery of Aircraft and Personnel (TRAP)
   (f) Seizure/Recovery of Selected Personnel or Material
   (g) Counter Proliferation (CP) of Weapons of Mass Destruction (WMD)

3. Military Operations Other Than War (MOOTW)
   (a) Peace Operations
      i) Peacekeeping
      ii) Peace Enforcement
   (b) Security Operations
      i) Non-combatant Evacuation Operations (NEO)
      ii) Reinforcement Operations
   (c) Joint/Combined Training/Instruction Team
   (d) Humanitarian Assistance/Disaster Relief

4. Supporting Operations
   (a) Tactical Deception Operations
   (b) Fire Support Planning, Coordination, and Control in a Joint/Combined Environment
   (c) Signal Intelligence (SIGINT)/Electronic Warfare (EW)
   (d) Military Operations in Urban Terrain (MOUT)
   (e) Reconnaissance and Surveillance (R&S)
   (f) Initial Terminal Guidance (ITG)
   (g) Counterintelligence Operations (CI)
   (h) Airfield/Port Seizure
   (i) Limited Expeditionary Airfield Operations
   (j) Show of Force Operations
   (k) JTF Enabling Operations
   (l) Sniping Operations

The MEU (SOC) has a limited:

1. Defensive capability against armored/motorized units in open terrain.
2. Defensive capability against a sustained low-level air attack when operating independent of naval air support.
3. Capability to replace combat losses and retrain if early introduction of follow-on forces is not contemplated.
4. Capability to participate in special warfare tasks requiring mobile training teams or nation-building efforts. However, the MEU (SOC) can provide some entry level and/or reinforcement training.
5. Ability to establish an MEU Headquarters ashore, and operate independent of Naval Shipping. The MEU (SOC) is heavily reliant upon shipboard facilities for C41 and aviation maintenance support.

The most significant difference between the current MEU (SOC) program and the old training for Marine Amphibious Units (MAU) in the 1970s/early 1980s is that an intense 26-week "work-up" exists. The training program is standardized and follows a progressive building-block approach to training. This training program integrates the Amphibious Squadron (PHIBRON) and the MEU as well as other designated forces (i.e., CVBG) to optimize coordination and use of capabilities. The 26-week "work-up" culminates in a Special Operations Capable Exercise (SOCEX) that realistically evaluates the MAU's warfighting capabilities. Only MEUs which have demonstrated proficiency in the skills and capabilities listed above will be designated as "MEU(SOC)."

SPECIAL PURPOSE MAGTF (SPMAGTF)
A Special-purpose MAGTF (SPMAGTF) is a non-standing MAGTF temporarily formed to conduct a specific mission. It is normally formed when a standing MAGTF is either inappropriate or unavailable. SPMAGTFs are organized, trained, and equipped to conduct a wide variety of missions ranging from crisis response, to regionally focused training exercises, to peacetime missions. Their SPMAGTF designation derives from the mission they are assigned, the location in which they will operate, or the name of the exercise in which they will participate (e.g., "SPMAGTF (X)," "SPMAGTF Somalia," "SPMAGTF UNITAS," "SPMAGTF Andrew," etc.).

Acronym	Element of
M	MISSION
E	ENEMY
T	TERRAINIWEATHER
T	TROOPS & FIRE SUPPORT AVAILABLE
T	TIME AVAILABLE

Estimate of the Situation

COMMANDER'S ESTIMATE

The commander's estimate of the situation is an orderly reasoning process by which a commander evaluates all significant factors affecting the situation. At higher levels of command, the commander may make his estimate based on the written or verbal estimates his staff develops. At these levels, it is customary for staff officers to assist the commander by:

- Assembling and interpreting information
- Making necessary assumptions
- Formulating possible courses of action
- Analyzing and comparing possible courses of action

METT-T

The commander uses the process represented by the acronym "METT-T" to analyze the situation. Each letter of the acronym stands for an important aspect of the situation.

Logistics and space: METT-T Commanders also consider the impact of logistics and any logistical requirements as well as the advantages and limitations of space.

NOTE: Logistics is usually considered as part of "troops and fire support available" (friendly combat power), while various aspects of *space* are considered under both "terrain" (maneuver area available) and time (the time/space relationship). Sometimes the acronym METT-TSL is used: Mission, Enemy, Terrain and Weather, Troops and Fire Support Available-Time-Space and Logistics.

Mission

The estimate of the situation is a systematic approach to problem solving. It helps the commander examine all relevant factors that affect his mission as well as select a feasible course of action that will accomplish his assigned responsibilities. The commander uses the acronym METT-T to analyze the mission and make his estimate.

- Analyzing your mission is one of five factors in the estimate of the situation.
- The mission is the task, together with its purpose, which clearly indicates the reason for the mission and the action to be taken.
- The mission must be carefully analyzed and understood within the context of the higher commander's intent. Specified and implied tasks are derived and assigned a relative priority.

- A mission clearly conveys the WHO, WHAT, WHEN, WHERE, and WHY of an operation. For example: *"AT H-Hour on D-day, BLT 1/5 attacks toward Objective A in order to fix the enemy in the vicinity of Hill 138."*

Enemy

The second factor to consider is the enemy situation. The commander considers all *of the enemy's capabilities. He makes plans to counteract or neutralize those enemy capabilities that can* prevent or hinder accomplishing his mission.

As the commander considers the enemy situation, he carefully distinguishes what he knows about the enemy from what he guesses or assumes about the enemy. He studies known dispositions, compositions, and strengths of the enemy in terms committed, reinforcing, and supporting forces. He attempts to "see" the enemy on the terrain. He mentally reviews recent significant activities, and he strives to find enemy weaknesses and peculiarities that can be exploited.

What are enemy capabilities? Enemy capabilities are:

"Those courses of action of which the enemy is physically capable, and that, if adopted, will affect accomplishment of our mission. The term 'capabilities' includes not only general courses of action open to the enemy, such as attack, defense, or withdrawal (DRAWD), but also all the particular courses of action possible under each general course of action. Enemy capabilities are considered in the light of all known factors affecting military operations, including time, space, weather, terrain, and the strength and disposition of enemy forces."

Determining the enemy commander's capabilities will become the key to any analysis of the enemy. It is not good enough to know numerous facts about the enemy unless you are capable of drawing conclusion as to his next action. It is only through this process that you are able to outmaneuver our opponent by anticipating his moves.

The commander determines what information he lacks. The successful commander is especially careful not to make arrogant assumptions about the enemy capabilities base upon race, religion, nationality, literacy, sophistication, sex, etc. Time and time again throughout history, military forces have suffered disastrous defeats because of overconfidence brought on by some false assumption of natural or national superiority.

Two important acronyms are used to communicate the enemy's strength, composition, weapons, combat efficiency, and capabilities. The acronyms SALUTE and DRAWD are the basis for further analysis of enemy capabilities. They are **tools** to assist in the evaluation and analysis of the enemy.

Use **SALUTE** to evaluate the enemy's known situation: *Unit Activity* time Location

Use **DRAWD** to evaluate enemy capabilities. Can the enemy: *Defend Withdraw* Reinforce *Delay Attack*

Troops and Fire Support Available

When commanders consider troops and the available fire support, they are developing their assessment of their **combat power.** Joint Publication 102 defines combat power as *"the total means of destructive and/or disruptive force which a military unit/formation can apply against the opponent at a given time."*

Let's look at this definition:

First of all, combat power is **always** relative. It is not an absolute. In this respect, combat power is much like maneuver in that its value has meaning only when considered in relation to the enemy's combat power.

Secondly, combat power is composed of both tangible and intangible factors. You must avoid the tendency to equate combat power with "bean counting" or "number crunching." For example, recall the old three-to-one advantage that the attacker should have to the defender: If the enemy has 200 men defending a position, you must have at least 600 troops to seize that position. As a former Commandant of the Marine Corps General Barrow said, *"Success in battle is not a function of how many show up, but who they are."*

If you use your imagination to generate superior combat power at a decisive time and place and are aggressive in employing it; you will win against otherwise seemingly very high odds.

This fourth factor of METT-T includes considerations of friendly capabilities. Some of the capabilities the commander will normally consider are:

- Equipment and weapons
- Tanks, infantry fighting vehicles, armored personnel carriers, machine guns, anti-tank missiles, and mortars
- Air support available
- Artillery support available
- Logistics capabilities
- Morale
- Training
- Leadership
- Fire support available

There are many assets for consideration when talking about fire support available as is mentioned above. Combined arms is the full integration of arms in such a way that in order to counteract one, the enemy must make himself more vulnerable to another. We pose the enemy not just with a problem, but with a dilemma, a no win situation. The Marine Corps takes advantage of the complementary characteristics of different types of units and enhances its mobility and firepower. For example, in order to defend against the infantry attack, the enemy must make himself vulnerable to the supporting arms. We use assault support to quickly concentrate superior ground forces for a breakthrough. We use artillery and close air support to support the infantry penetration, and we use deep air support to interdict enemy reinforcements. The plan of attack must consist of a scheme of maneuver and a fire support plan. These two aspects must complement each other. For example, at the very lowest level is the complementary use of the automatic weapon and grenade launcher within a fire team. We pin an enemy down with the high-volume, direct fire of the automatic weapon, making him a vulnerable target for the grenade launcher. If he moves to escape the impact of the grenades, we engage him with the automatic weapon.

Troops -Within the "Troops and Fire Support Available" portion of the estimate process, you must consider logistics. Joint Publication 102 defines logistics as *"the science of planning and carrying out the movement and maintenance of forces."*

One description of the role logistics plays in the commander's estimate comes from British General Archibald Wavell:

"The more I see of war, the more I realize how it all depends on administration and transportation . . . It takes little skill or imagination to see where you would like your army to be and when; it takes much knowledge and hard work to know where you can place your forces and whether you can maintain them there."

A real knowledge of supply and movement factors must be the basis of every leader's plan. Only then can the leader know how and when to take risks with those factors. The key point in the quotation above is:

"it takes much knowledge and hard work to know where you can place your forces and whether you can maintain them there."

Terrain and Weather

The fourth factor of the estimate of the situation is to analyze the terrain that you will operate in and the weather conditions. **COKOA** is a mnemonic device used for identifying terrain information which the leader derives from a personal reconnaissance or through a careful map study. COKOA stands for:

- Cover and concealment
- Obstacles
- Key terrain
- Observation and fields of fire
- Avenues of approach

An obstacle is anything, including a natural or artificial terrain feature, that stops, impedes, or diverts military movement. Obstacles may either be existing or reinforcing.

The mission influences the determination of obstacles. In the attack, the commander considers the features within his unit's zone of action. In the commander's opinion, an obstacle may be an advantage or disadvantage.

Obstacles perpendicular to the direction of attack favor the defender by slowing or canalizing the attacker. Obstacles parallel to the direction of attack may assist in protecting a flank of the attacking force.

To fully understand what constitutes an obstacle, commanders must first consider their mission and means of mobility.

An **avenue of approach** is: *"An air or ground route of an attacking force of a given size leading to its objective or to key terrain in its path."*

What is really meant by an avenue of approach? An avenue of approach is a route by which a force may reach its objective. In planning for use of avenues of approach you should consider:

All of the other factors of key terrain, observation and fire, concealment and cover, and obstacles, from both friendly and enemy perspectives.

The mission, type, and size of the unit.

Generally, commanders must consider avenues of approach that are adequate for the forces operating in the maneuver space. Intelligence about each avenue is checked against possible schemes of maneuver, both friendly and enemy. Maneuver space should be adequate. This consideration is based on deployment patterns and the means of movement. Ease of movement is considered and includes such factors as:

- Soil type
- Road trafficability
- River trafficability
- Steepness of slopes
- Terrain compartments
- Vegetation

A key terrain feature is: *"Any locality or area the seizure or retention of which affords a marked advantage to either combatant."*

Based on the mission of the command, commanders must consider key terrain features in formulating courses of action.

Terrain which permits or denies maneuver may be key terrain.

Tactical use of terrain often emphasizes increasing the ability to apply friendly combat power at the same time forcing the enemy into areas which reduce his combat power. Terrain which permits this may also be key terrain. Considerations in selecting key terrain are:

- The effect of terrain on fire and maneuver
- The application of combat power
- The preservation of force integrity

A terrain feature may afford a marked advantage in one set of circumstances but little or no advantage under other conditions. Selecting key terrain varies with the level of command, the type of unit, and the mission of the unit.

Combat service support (CSS) and aviation units need key terrain:

- CSS units need roads over which to move supplies and secure areas in which to establish facilities.
- Aviation units need high terrain on which to set up radar and communication facilities and large flat areas for airfields.

Observation and fields of fire are so closely related that they are considered together. They are not synonymous, but fields of fire are based on observation because troops must see a target to bring effective fire upon it.

"Observation is the area over which surveillance can be exercised either visually or through the use of surveillance devices, both optical and electronic."

A field of fire is: *"The area a weapon or group of weapons may cover effectively with fire from a given position."*

- Observation varies with
- Weather conditions
- The time of day
- Vegetation
- Friendly and enemy smoke
- Surrounding terrain

Observation generally is best from the highest terrain features. However, during times of poor visibility, positions in low areas that the enemy must pass through may provide better observation than high points from which nothing can be seen.

Fields of fire for direct fire weapons, such as machine guns and automatic rifles, may be affected by terrain conditions between the weapon and the target.

A leader identifies those terrain features within the area of operations and those areas adjacent to the area of operations. These are terrain features which afford the friendly or enemy force favorable observation and fire.

Concealment and cover are protection from observation and fire. Specifically, concealment is *"the protection from observation or surveillance."* *"Cover is protection from the effects of fire."*

What is good concealment? Woods, underbrush, snowdrifts, tall grasses, cultivated vegetation, or any other feature which denies observation, usually provide good concealment. Good concealment may also be provided by weather conditions, such as fog and rain, and by darkness. Concealment from ground observation does not necessarily provide concealment from air observation or from electronic or infrared detection devices.

Remember, terrain that provides concealment **may or may not** provide cover!

Cover may be provided by rocks, ditches, quarries, caves, river banks, folds in the ground, shell craters, buildings, walls, railroad embankments and cuts, sunken roads, and highway fills. Areas that provide cover from direct fire may or may not protect against the effects of indirect fire. Most terrain features that offer cover also provide concealment from ground observation.

Although you have studied the military aspects of terrain and weather separately, terrain and weather are, in fact, inseparable. For example, terrain that offers good trafficability when it is dry may be impassable when it is wet. A hill that offers good observation on a clear day may not provide any visibility on a rainy day or during foggy conditions. Now, let's turn to our next factor in the estimate of the situation.

"Weather is the state of the atmosphere at a given time and place. It includes atmospheric pressure, winds, humidity, clouds and fog, precipitation, and fronts or zones where air masses of different temperatures meet."

What does all that mean to a military force? Let's consider this quote from General Eisenhower's book, *Crusade in Europe:*

"Some soldier once said 'The weather is always neutral.' Nothing could be more untrue. Bad weather is obviously the enemy of the side that seeks to launch projects requiring good weather, or of the side possessing great assets, such as strong air forces, which depend upon good weather for effective operation."

Eisenhower means that a certain type of weather in and of itself is not always good or bad. Whether the weather is good or bad for an operation depends on the requirements of a specific operation, conducted in response to a specific situation. The key point is that what U.S. forces traditionally refer to as "bad weather" can be a great ally.

Time

Time is the commander's most important resource! What is meant by the term time available? Perhaps the best and clearest comment about the role of time in the offense is Clausewitz's injunction: *"Time which is allowed to pass unused accumulates to the credit of the defender."*

Time is closely tied to the concepts of **momentum and culminating point.** If the attacker is to succeed, he must constantly concentrate on imaginatively and aggressively getting the most from each moment. Time is closely linked to space.

Make a time plan. The goal is to give the subordinate unit enough daylight to conduct planning, reconnaissance, and preparation before the start of combat operations. It does more harm than good to present a perfect plan to subordinate units if they do not have the time to disseminate their own orders and prepare. Plan your time by using the "half-rule" or the "one-third, two-thirds rule."(For example, half of the available time goes to the commander and half goes to the subordinate units.)

Backward planning is also necessary to ensure an effective use of time. Start with the last known action and progress backward to present time. This will be the time for crossing of the line of departure for an offensive battle or from the time the defense must be established.

The commander must consider the distance he must move in the required time. This is why time and space are considered together. The commander should compute:

- How much time will be needed to move certain distances
- How far from the objective he must begin to change formations to begin the assault

Compute this time/distance estimate with regards to specific conditions, such as weather or the enemy situation. You must anticipate friction, such as obstacles or harassing fire from the enemy that your troops may encounter. It will slow down friendly units.

REVIEW QUESTIONS

1. What is the goal of Maneuver Warfare?
2. What is the basis for the Marine Corps Philosophy of Command?
3. What are the three phases of the offense?
4. Just before the Final Coordination Line the assault element is in what position?
5. Define: METT-T, DRAW-D, COKOA, SALUTE.
6. What is combined arms?

CHAPTER 13

Strike Warfare

LEARNING OBJECTIVES

At the end of this chapter the student will be able to:

- Define Strike Warfare. State the command relationship between the CWC and the STWC.
- Describe the four mission types of Strike Warfare.
- Describe the various platforms and weapons involved is STW.
- Describe the major components of the Tomahawk Weapon System.
- Describe the advantages and disadvantages of Tomahawk and TACAIR.

ADDITIONAL READING

Friedman, Norman. *World Naval Weapon Systems.* Annapolis, MD.: Naval Institute Press, 1989.

Giauque, Michael, LT. "Cruising Ahead with Tomahawk." *Warfare Magazine* (September/October 1992): 811.

Polmar, Norman. *The Ships and Aircraft of the U.S. Fleet* (14th ed.). Annapolis, MD.: Naval Institute Press, 1987.

Prezelin, Bernard and Baker A. D. *Combat Fleets of the World 19901–1991, Their Ships, Aircraft, and Armament.* Annapolis: Naval Institute Press, 1990.

Worldwide Challenges to Strike Warfare, Office of Naval Intelligence, January 1996.

Worldwide Challenges to Strike Warfare, Office of Naval Intelligence, February 1997.

Introduction

Strike Warfare is the use of tactical aircraft and/or cruise missile strikes against land targets in an offensive power projection role. During World War II, aircraft gave birth to the concept of naval strike warfare. The ability of aircraft to appear unexpectedly, anywhere within the battlespace, made them particularly effective. Today, the value of naval strike warfare has increased significantly with the addition of precision-guided bombs and cruise missiles.

Strike Warfare is characterized by four missions: coordinated strike, interdiction, armed reconnaissance, and close air support. The Strike Warfare Commander (STWC) will select one of these mission types whenever a strike is ordered. The selection of mission type is dependent upon the goals of the strike. The STWC is a subordinate commander under the Composite Warfare Commander (CWC) and is specifically trained in all aspects of Tactical Air (TACAIR) and Tomahawk strike capabilities.

Coordinated Strike

The coordinated strike mission (also termed deep tactical support) is to destroy specified targets at known locations. The objective of Coordinated Strike is to reduce the enemy's war-making capacity and logistic capability. Targets for this mission are likely to be well inland and highly defended. Strikes may require relatively large numbers of support aircraft to protect and assist strike aircraft. Large numbers of strike aircraft may be needed to attain reasonable levels of destruction in order to avoid having to conduct a re-strike operation.

Interdiction

The interdiction mission is to destroy specifically briefed targets which deny the enemy access to an area. It has a secondary mission of attacking targets of opportunity, if so authorized.

Armed Reconnaissance

The armed reconnaissance mission is to destroy targets of opportunity. It has a secondary mission of attacking specified fixed targets, if no target of opportunity presents itself. Armed reconnaissance is generally planned for a specified route or area, and weapons loads may be tailored to attack moving or movable targets.

Close Air Support

The close air support (CAS) mission is to harass, neutralize, or destroy enemy ground forces that present an immediate or direct threat to friendly ground forces. Close air support is provided in accordance with the supporting commander's tactical requirements and ability to control or coordinated the effort.

STW Platforms and Weapons

ATTACK AIRCRAFT

The Navy's premier strike fighter is the F/A-18 E/F SUPER HORNET. It incorporates a number of improvements over F/A18 C/D including:

- 25 percent larger wing
- New engines
- Up to 40 percent more range
- Updated cockpit
- Additional weapons stations
- Reduced radar signature
- Increased survivability
- Growth potential

STRIKE WARFARE

The F/A-18 E/F is able to employ a wide variety of precision-guided weapons including laser-guided bombs and the new family of joint attack weapons (JDAM and JSOW). This important strike capability is achieved without compromising air-to-air performance.

AIR-LAUNCHED WEAPONS

Precision-guided weapons like the Joint Stand-Off Weapon (JSOW), Joint Directed Attack Munition (JDAM), Stand-Off Land Attack Missile Expanded Response (SLAM-ER), laser-guided bombs, and the TOMAHAWK cruise missile make up the lethal punch of Naval Strike Warfare.

The **JOINT STAND-OFF WEAPON (JSOW)** is a joint development effort by the Navy and Air Force to produce the next generation of stand-off missiles. The Navy is the lead Service in developing the weapon. The JSOW is a subsonic weapon with an approximate range of 40 nm. This missile can be armed with a 500–1,000 –pound warhead, BLU-108 bomblets, or Brilliant AntiTank (BAT) submunition. The JSOW is a GPS-guided glide bomb.

The **Joint Directed Attack Munition (JDAM)** is a guided air-to-surface weapon that uses either the 2,000-pound BLU-109/MK 84, the 1,000-pound BLU-110/MK 83, or the 500-pound BLU-111/MK 82 warhead as the payload. JDAM enables employment of accurate air-to-surface weapons against high priority fixed and relocatable targets from fighter and bomber aircraft. Guidance is facilitated through a tail control system and a GPS-aided INS. The navigation system is initialized by transfer alignment from the aircraft that provides position and velocity vectors from the aircraft systems. Once released from the aircraft, the JDAM autonomously navigates to the designated target coordinates. Target coordinates can be loaded into the aircraft before takeoff, manually altered by the aircrew before weapon release, or automatically entered through target designation with onboard aircraft sensors. In its most accurate mode, the JDAM system will provide a weapon circular error probable of 5 meters or less during free flight when GPS data is available. If GPS data is denied, the JDAM will achieve a 30-meter CEP or less for free flight times up to 100 seconds with a GPS quality handoff from the aircraft.

JDAM can be launched from very low to very high altitudes in a dive, toss, or loft and in straight and level flight with an on-axis or off-axis delivery. JDAM enables multiple weapons to be directed against single or multiple targets on a single pass. JDAM is currently compatible with B-1B, B-2A, B-52H, AV-8B, F-15E, F/A 18C/D/E/F, F-16C/D and F-22 aircraft. Follow-on integration efforts are currently under way or planned to evaluate compatibility with the A-10, F-35 Joint Strike Fighter, and MQ-9 Reaper unmanned aerial vehicle. JDAM is a joint U.S. Air Force and Department of Navy program.

The **STAND-OFF LAND ATTACK MISSILE-EXPANDED RESPONSE (SLAM-ER)** uses a Harpoon's propulsion section and warhead. The SLAM-ER is a subsonic missile with an approximate range of 120 nm. U.S. Navy and Marine Corps Aircraft carry Mk 80 series bombs. The Mk 82 (500 pound), Mk 83 (1,000 pound), and Mk 84 (2,000 pound) bombs can be outfitted with laser-guided bomb kits enabling launching aircraft or other forces to guide a bomb to a target illuminated by a laser designator. Laser-guided bombs require aircrews to locate and designate the target throughout weapon flight.

CRUISE MISSILES

TLAM-C (Tomahawk Land Attack Missile–Conventional) is a 1,000 –pound warhead installed. TLAM-C has a range of about 675 nm. Guidance for TLAM C and D are the same, both use inertial navigation and TERCOM to contour match the surface land below. Additionally, DSMAC is used to compare scenes optically viewed to an onboard digital map. Block III missiles also have GPS capability. (For an explanation of TLAM terms, see below.)

TLAM-D (Tomahawk Land Attack Missile–Bomblet Dispenser). The bomblets are numerous small-shaped charges used to cover an area or several different targets. TLAM-D has a range of about 475 nm.

TLAM-N (Tomahawk Land Attack Missile-Nuclear) is a nuclear warhead (W80 with 200-kT) attached to the Tomahawk airframe. TLAM-N has an approximate range of 1300 nm.

Tomahawk Components

DTD (Data Transport Device): A storage device of mission information that is loaded into the Tomahawk missile upon mission selection prior to launch. After launch the mission can not be updated.

GPS (Global Positioning Information): A satellite-based positioning system. The same positioning system used in electronic navigation is used in conjunction with TERCOM and/or DSMAC in the guidance of a Tomahawk cruise missile. The GPS encryption for the "P" code is loaded into the Tomahawk during missile alignment.

APS (Afloat Planning System): A deployable detachment of Tomahawk mission planners and equipment that can develop, revise, and disseminate Tomahawk missions within the CV Battle Group's area of operation.

TERCOM (Terrain Contour Matching): A method to navigate the weapon by comparing the Tomahawk's inertial position and altitude with a contour map stored in the missile's onboard computer.

DSMAC (Digital Scene Matching Area Correlation): Select digital scenic photographs leading to the target stored in the mission software. The missile uses an onboard camera to compare photos with scenes stored in the software. This is where precise missile navigation occurs.

WARHEAD: 1,000 -pound warhead, bomblet package, or nuclear warhead.

MDU: (Mission Data Update). Missions transferred to ships via satellites and loaded into existing DTDs. Provides short term mission planning flexibility.

TAINS: (TERCOM Aided Inertial Navigation). Basically a dead reckoning system with TERCOM updates.

The figure to the right shows a Tomahawk missile receiving preplanned mission information from shore planning facilities via the launch platform.

When the missile arrives at the first preplanned waypoint, it checks its navigation position through a series of TERCOM checks. As the Tomahawk approaches the target, very accurate navigation is achieved through a series of DSMAC maps.

Planning Considerations

The Strike Warfare Commander must comprehend the advantages and disadvantage of using Tomahawk vs. TACAIR and consider other aspects when preparing a "Strike Package."

Platforms and Planning Considerations

Targeting and Weaponeering

Intelligence and Strike Warfare are inextricably linked. Intel provides crucial data for the targeting and weaponeering processes.

Targeting: The selection of targets to be struck as part of the campaign or strategy. Closely related to concepts of "critical vulnerability" and "center of gravity."

Examples: Targeting of radio relay stations to degrade enemy's C2 capability. Targeting a party headquarters to damage a country's leadership. Targeting an airfield to degrade air defense capability.

Weaponeering: The pairing of weapons with targets to ensure maximum effectiveness and minimum risk. Examples: Using TLAM against SAM sites that present a high risk to manned aircraft. Using F/A-18s with LGBs against mobile units that are difficult for TLANls to hit.

Time-Critical Targets are a growing problem. Intel and strike planners must work closely together to provide data for strike weapons (AIC or TLAM) in order to hit mobile targets such as SAM sites that redeploy daily in southern Iraq.

Campaigns are almost always joint conducted under the command of the theater CINCo He designates a JFACC, who oversees all air operations, including strike operations. As part of that, his staff conducts a target prioritization process, which works as follows:

The ATO Process

Component commanders (JFMCC, JFLCC, JFACC, etc.) submit a target "wish list" to the GAT (Guidance, Apportionment, Targeting) Cell. The GAT cell merges those lists and creates a Joint Prioritized Integrated Target List (JPITL).

The component commanders also submit a list of available strike assets. Those assets are matched with the **JPITL** to create the Air Tasking Order (ATO).

TLAM targeting is also conducted at the theater CINC level. Once a target is designated for TLAM, a Cruise Missile Support Activity or an Afloat Planning System detachment plans the mission. Once complete the mission is transmitted to a TLAM shooter for execution.

REVIEW QUESTIONS

1. What is the definition of Strike Warfare?
2. What are the four types of strike mission?
3. What is primary tactical aircraft used in strike missions?
4. What are the different versions of Tomahawk?
5. What is an MDU?
6. Generally, how does a Tomahawk missile find its way to the target?
7. What are the advantages and disadvantages of using strike aircraft versus Tomahawk?
8. What are some of the planning considerations when TACAIR is selected?

Appendix: Legislative Changes to the JCS

Legislation	Provisions
1947 National Security Act	<ul><li>Designated Secretary of National Defense to exercise **general** authority, direction, and control</li><li>Created the **National Military Establishment**</li><li>Established U.S. Air Force</li><li>Established CIA and NSC</li><li>Established JCS as permanent agency</li><li>JCS became **principal military advisers** to President and Secretary of Defense</li><li>Established a legal basis for unified and specified commands</li></ul>
1948 Key West Agreement	<ul><li>Established JCS as **executive agents** for unified and specified commands</li><li>Service roles and missions defined</li></ul>
1948 Amendment	<ul><li>Military department Secretaries reduced from cabinet rank and removed from NSC</li><li>Renamed NME the Department of Dfense</li><li>Created office of **Chairman**</li></ul>
1952 Amendment	<ul><li>Gave **Commandant of the Marine Corps** (CMC) co-equal status on JCS on Marine Corps issues</li></ul>
1953 Plan	<ul><li>**Removed JCS from executive agent status**, i.e., handling day-to-day communications and supervision over unified commands</li><li>Established **military departments as executive agents** for unified commands</li></ul>
1958 Amendment	<ul><li>Gave Chairman a vote</li><li>**Removed military departments as executive agents**</li><li>Joint Staff has no executive authority, but assists the Secretary of Defense in exercising direction over unified commands</li></ul>
1978 Amendment	<ul><li>Made CMC a full member of JCS</li></ul>
1986 Amendment	<ul><li>Designated Chairman **principal military adviser**</li><li>Transferred duties of corporate JCS to Chairman</li><li>Created position of Vice Chairman</li><li>Specified chain of command to run from President to Secretary of Defense to unified and specified combatant commanders</li></ul>

References: National Security Act of 1947, as amended; Reorganization of the National Security Organization, Report of the CNO Select Panel, dated March 1985.

Glossary

administrative control (ADCON). Direction or exercise of authority over subordinate or other organizations in respect to administration and support, including organization of Service forces, control of resources and equipment, personnel management, unit logistics, individual and unit training, readiness, mobilization, demobilization, discipline, and other matters not included in the operational missions of the subordinate or other organizations. (JP 1-02. Source: JP 1)

area of operations (AO). An operational area defined by the joint force commander for land and maritime forces. Areas of operation do not typically encompass the entire operational area of the joint force commander, but should be large enough for component commanders to accomplish their missions and protect their forces. (JP 1-02. Source: JP 3-0)

assessment. 1. A continuous process that measures the overall effectiveness of employing joint force capabilities during military operations. 2. Determination of the progress toward accomplishing a task, creating an effect, or achieving an objective. (JP 1-02. Source: JP 3-0)

attach. The placement of units or personnel in an organization where such placement is relatively temporary. (JP 1-02. Source: JP 3-0)

battle rhythm. A deliberate daily cycle of command, staff, and unit activities intended to synchronize current and future operations. (JP 1-02. Source: JP 3-33)

carrier strike group (CSG). The combining of Navy, naval, and perhaps other maritime capabilities that provides the full range of operational capabilities for sustained maritime power projection and combat survivability. The baseline organization consists of a carrier strike group command element/ staff, a destroyer squadron command element/staff, one aircraft carrier, one carrier air wing, five surface combatant ships, one cruise missile land attack/undersea warfare submarine (SSN), one or two multi-product logistic support ship, and one logistics helicopter detachment. (OPNAVINST 3501.316A)

close support. That action of the supporting force against targets or objectives which are sufficiently near the supported force as to require detailed integration or coordination of the supporting action with the fire, movement, or other actions of the supported force. (JP 1-02. Source: JP 3-31)

combatant command (command authority) (COCOM). Nontransferable command authority established by title 10 ("Armed Forces"), United States Code, section 164, exercised only by commanders of unified or specified combatant commands unless otherwise directed by the President or the Secretary of Defense. Combatant command (command authority) cannot be delegated and is the authority of a combatant commander to perform those functions of command over assigned forces involving organizing and employing commands and forces, assigning tasks, designating objectives, and giving authoritative direction over all aspects of military operations, joint training, and logistics necessary to accomplish the missions assigned to the command. Combatant command (command authority) should be exercised through the commanders of subordinate organizations. Normally this authority is exercised through subordinate joint force commanders and Service and/or functional component commanders. Combatant command (command authority) provides full authority to organize and employ commands and forces as the combatant commander considers necessary to accomplish assigned missions. Operational control is inherent in combatant command (command authority). (JP 1-02. Source: JP 1)

command and control (C2). The exercise of authority and direction by a properly designated commander over assigned and attached forces in the accomplishment of the mission. Command and control functions are performed through an arrangement of personnel, equipment, communications, facilities, and procedures employed or other actions of the supported force by a commander in planning, directing, coordinating, and controlling forces and operations in the accomplishment of the mission. (JP 1-02. Source: JP 1)

command relationships. The interrelated responsibilities between commanders, as well as the operational authority exercised by commanders in the chain of command; defined further as combatant command (command authority), operational control, tactical control, or support. (JP 1-02. Source: JP 1)

commander's critical information requirement (CCIR). An information requirement identified by the commander as being critical to facilitating timely decision making. The two key elements are friendly force information requirements and priority intelligence requirements. (JP 1-02. Source: JP 3-0)

commander, Navy forces (COMNAVFOR). The senior Navy commander assigned to a joint task force which does not have the Navy component commander assigned to it. (NWP 3-32)

composite warfare commander (CWC). The officer in tactical command is normally the composite warfare commander. However the composite warfare commander concept allows an officer in tactical command to delegate tactical command to the composite warfare commander. The composite warfare commander wages combat operations to counter threats to the force and to maintain tactical sea control with assets assigned, while the officer in tactical command retains close control of power projection and strategic sea control operations. (JP 1-02. Source: JP 3-02)

coordinating authority. A commander or individual assigned responsibility for coordinating specific functions or activities involving forces of two or more Military Departments, two or more joint force components, or two or more forces of the same Service. The commander or individual has the authority to require consultation between the agencies involved, but does not have the authority to compel agreement. In the event that essential agreement cannot be obtained, the matter shall be referred to the appointing authority. Coordinating authority is a consultation relationship, not an authority through which command may be exercised. Coordinating authority is more applicable to planning and similar activities than to operations. (JP 1-02. Source: JP 1)

critical capability. A means that is considered a crucial enabler for a center of gravity to function as such and is essential to the accomplishment of the specified or assumed objective(s). (JP 1-02. Source: JP 5-0)

critical requirement. An essential condition, resource, and means for a critical capability to be fully operational. (JP 1-02. Source: JP 5-0)

critical vulnerability. An aspect of a critical requirement which is deficient or vulnerable to direct or indirect attack that will create decisive or significant effects. (JP 1-02. Source: JP 5-0)

culminating point. The point at which a force no longer has the capability to continue its form of operations, offense or defense. (JP 1-02. Source: JP 5-0)

decisive point. A geographic place, specific key event, critical factor, or function that when acted upon, allows commanders to gain a marked advantage over an adversary or contribute materially to achieving success. (JP 1-02. Source: JP 3-0)

direct liaison authorized (DIRLAUTH). That authority granted by a commander (any level) to a subordinate to directly consult or coordinate an action with a command or agency within or outside of the granting command. Direct liaison authorized is more applicable to planning than operations and always carries with it the requirement of keeping the commander granting direct liaison authorized informed. Direct liaison authorized is a coordination relationship, not an authority through which command may be exercised. (JP 1-02. Source: JP 1)

direct support (DS). A mission requiring a force to support another specific force and authorizing it to answer directly to the supported force's request for assistance. (JP 1-02. Source: JP 3-09.1)

expeditionary strike group (ESG). The combining of Navy, naval, and perhaps other maritime capabilities that provides the organic air defense, expeditionary warfare capability, and strike capability required for operating independently in low-to-medium threat environments. The baseline organization consists of a expeditionary strike group command element (when required), one amphibious squadron command element/staff, three amphibious ships, one Marine expeditionary unit or Marine expeditionary unit (special operations capable), three surface combatant ships, and one cruise missile land attack/undersea warfare submarine (SSN) (when required). (OPNAVINST 3501.316A)

fires. The use of weapon systems to create a specific lethal or nonlethal effect on a target. (JP 1-02. Source: JP 3-0)

fleet. An organization of ships, aircraft, Marine forces, and shore-based fleet activities all under the command of a commander who may exercise operational as well as administrative control. (JP 1-02. Source: N/A)

friendly force information requirement (FFIR). Information the commander and staff need to understand the status of friendly force and supporting capabilities. (JP 1-02. Source: JP 3-0)

full command. The command authority which covers every aspect of military operations and administration that exists only within national services and, therefore, is always retained by national commanders. (NWP 3-32)

general support (GS). That support which is given to the supported force as a whole and not to any particular subdivision thereof. (JP 1-02. Source: N/A)

joint force air component commander (JFACC). The commander within a unified command, subordinate unified command, or joint task force responsible to the establishing commander for making recommendations on the proper employment of assigned, attached, and/or made available for tasking air forces; planning and coordinating air operations; or accomplishing such operational missions as may be assigned. The joint force air component commander is given the authority necessary to accomplish missions and tasks assigned by the establishing commander. (JP 1-02. Source: JP 3-0)

joint force commander (JFC). A general term applied to a combatant commander, subunified commander, or joint task force commander authorized to exercise combatant command (command authority) or operational control over a joint force. (JP 1-02. Source: JP 1)

joint force land component commander (JFLCC). The commander within a unified command, subordinate unified command, or joint task force responsible to the establishing commander for making recommendations on the proper employment of assigned, attached, and/or made available for tasking land forces; planning and coordinating land operations; or accomplishing such operational missions as may be assigned. The joint force land component commander is given the authority necessary to accomplish missions and tasks assigned by the establishing commander. (JP 1-02. Source: JP 3-0)

joint force maritime component commander (JFMCC). The commander within a unified command, subordinate unified command, or joint task force responsible to the establishing commander for making recommendations on the proper employment of assigned, attached, and/or made available for tasking maritime forces and assets; planning and coordinating maritime operations; or accomplishing such operational missions as may be assigned. The joint force maritime component commander is given the authority necessary to accomplish missions and tasks assigned by the establishing commander. (JP 1-02. Source: JP 3-0)

joint operations area (JOA). An area of land, sea, and airspace, defined by a geographic combatant commander or subordinate unified commander, in which a joint force commander (normally a joint task force commander) conducts military operations to accomplish a specific mission. (JP 1-02. Source: JP 3-0)

joint task force (JTF). A joint force that is constituted and so designated by the Secretary of Defense, a combatant commander, a subunified commander, or an existing joint task force commander. (JP 1-02. Source: JP 1)

leverage. In the context of joint operation planning, a relative advantage in combat power and/or other circumstances against the adversary across one or more domains (air, land, maritime, and space) and/or the information environment sufficient to exploit that advantage. Leverage is an element of operational design. (JP 1- 02. Source: JP 5-0)

line of operations (LOO). 1. A logical line that connects actions on nodes and/or decisive points related in time and purpose with an objective(s). 2. A physical line that defines the interior or exterior orientation of the force in relation to the enemy or that connects actions on nodes and/or decisive points related in time and space to an objective(s). (JP 1-02. Source: JP 3-0)

logistics. Planning and executing the movement and support of forces. It includes those aspects of military operations that deal with: a. design and development, acquisition, storage, movement, distribution, maintenance, evacuation, and disposition of materiel; b. movement, evacuation, and hospitalization of personnel; c. acquisition or construction, maintenance, operation, and disposition of facilities; and d. acquisition or furnishing of services. (JP 1-02. Source: JP 4-0)

made available. Forces assigned and/or attached to a joint force service or functional component commander. (NWP 3-32)

major fleet. A principal, permanent subdivision of the operating forces of the Navy with certain supporting shore activities. Presently there are two such fleets: the Pacific Fleet and the Atlantic Fleet. See also fleet. (JP 1- 02. Source: N/A)

Marine air-ground task force (MAGTF). The Marine Corps principal organization for all missions across the range of military operations, composed of forces task-organized under a single commander capable of responding rapidly to a contingency anywhere in the world. The types of forces in the Marine air-ground task force (MAGTF) are functionally grouped into four core elements: a command element, an aviation combat element, a ground combat element, and a combat service support element. The four core elements are categories of forces, not formal commands. The basic structure of the MAGTF never varies, though the number, size, and type of Marine Corps units comprising each of its four elements will always be mission dependent. The flexibility of the organizational structure allows for one or more subordinate MAGTFs to be assigned. (JP 1-02. Source: N/A)

Marine expeditionary brigade (MEB). A Marine air-ground task force that is constructed around a reinforced infantry regiment, a composite Marine aircraft group, and a combat logistics regiment. The Marine expeditionary brigade (MEB), commanded by a general officer, is task-organized to meet the requirements of a specific situation. It can function as part of a joint task force, as the lead echelon of the Marine expeditionary force (MEF), or alone. It varies in size and composition and is larger than a Marine expeditionary unit but smaller than a MEF. The MEB is capable of conducting missions across the full range of military operations. (JP 1-02. Source: JP 3-18)

Marine expeditionary force (MEF). The largest Marine air-ground task force (MAGTF) and the Marine Corps principal warfighting organization, particularly for larger crises or contingencies. It is task-organized around a permanent command element and normally contains one or more Marine divisions, Marine aircraft wings, and Marine force service support groups. The Marine expeditionary force is capable of missions across the range of military operations, including amphibious assault and sustained operations ashore in any environment. It can operate from a sea base, a land base, or both. (JP 1-02. Source: N/A)

Marine expeditionary unit (MEU). A Marine air-ground task force (MAGTF) that is constructed around an infantry battalion reinforced, a helicopter squadron reinforced, and a task-organized combat service support element. It normally fulfills Marine Corps forward sea-based deployment requirements. The Marine expeditionary unit provides an immediate reaction capability for crisis response and is capable of limited combat operations. (JP 1-02. Source: N/A)

maritime domain. The oceans, seas, bays, estuaries, islands, coastal areas, and the airspace above these, including the littorals. (JP 1-02 Source: JP 3-32)

maritime headquarters (MHQ). A naval headquarters at the operational level supporting operational and administrative chains of command. (NWP 3-32)

maritime operations center (MOC). 1. The collective name for the boards, bureaus, cells, centers, and working groups that execute the maritime headquarters maritime operations functions. 2. A physical space in the maritime headquarters that is principally used for the monitoring, assessing, planning, and direction of current operations. (NWP 3-32)

maritime superiority. That degree of dominance of one force over another that permits the conduct of maritime operations by the former and its related land, maritime, and air forces at a given time and place without prohibitive interference by the opposing force. (JP 1-02. Source: JP 3-32 CH 1)

measure of effectiveness (MOE). A criterion used to assess changes in system behavior, capability, or operational environment that is tied to measuring the attainment of an end state, achievement of an objective, or creation of an effect. (JP 1-02. Source: JP 3-0)

measure of performance (MOP). A criterion used to assess friendly actions that is tied to measuring task accomplishment. (JP 1-02. Source: JP 3-0)

mission type order. An order to a unit to perform a mission without specifying how it is to be accomplished. (JP 1-02. Source: JP 3-50)

mutual support. That support which units render each other against an enemy, because of their assigned tasks, their position relative to each other and to the enemy, and their inherent capabilities. (JP 1-02. Source: N/A)

numbered fleet. A major tactical unit of the Navy immediately subordinate to a major fleet command and comprising various task forces, elements, groups, and units for the purpose of prosecuting specific naval operations. (JP 1-02. Source: N/A)

officer in tactical command (OTC). In maritime usage, the senior officer present eligible to assume command, or the officer to whom the senior officer has delegated tactical command. (JP 1-02. Source: N/A)

operational art. The application of creative imagination by commanders and staffs—supported by their skill, knowledge, and experience—to design strategies, campaigns, and major operations and organize and employ military forces. Operational art integrates ends, ways, and means across the levels of war. (JP 1-02. Source: JP 3-0)

operational command (OPCOM). A command authority granted to an allied/multinational maritime commander by a national commander with full command to assign missions or tasks to subordinated commanders, to deploy units, to reassign forces, and to retain or delegate operational control, tactical command, or tactical control as may be deemed necessary. It does not in itself include administrative command or logistical responsibility. (NWP 3-32)

operational control (OPCON). 1. Command authority that may be exercised by commanders at any echelon at or below the level of combatant command. Operational control is inherent in combatant command (command authority) and may be delegated within the command. Operational control is the authority to perform those functions of command over subordinate forces involving organizing and employing commands and forces, assigning tasks, designating objectives, and giving authoritative direction necessary to accomplish the mission. Operational control includes authoritative direction over all aspects of military operations and joint training necessary to accomplish missions assigned to the command. Operational control should be exercised through the commanders of subordinate organizations. Normally this authority is exercised through subordinate joint force commanders and Service and/or functional component commanders. Operational control normally provides full authority to organize commands and forces and to employ those forces as the commander in operational control considers necessary to accomplish assigned missions; it does not, in and of itself, include authoritative direction for logistics or matters of administration, discipline, internal organization, or unit training. (JP 1-02. Source: JP 1) 2. A command authority granted to an allied/multinational maritime commander by a national commander with full command or an allied/multinational maritime commander with operational command to direct forces assigned so that the commander can accomplish specific missions or tasks which are usually limited by function, time, or location; to deploy units concerned; and to retain or assign tactical command and/or control of those units. It does not include the authority to assign separate employment of the units concerned. Neither does it, of itself, include administrative command or logistic responsibility. Subordinate to operational command. (NWP 3-32)

operational design. The conception and construction of the framework that underpins a campaign or major operation plan and its subsequent execution. (JP 1-02. Source: JP 3-0)

operational intelligence. Intelligence that is required for planning and conducting campaigns and major operations to accomplish strategic objectives within theaters or operational areas. (JP 1-02. Source: JP 2-0)

operational level of war (OLW). The level of war at which campaigns and major operations are planned, conducted, and sustained to achieve strategic objectives within theaters or other operational areas. Activities at this level link tactics and strategy by establishing operational objectives needed to achieve the strategic objectives, sequencing events to achieve the operational objectives, initiating actions, and applying resources to bring about and sustain these events. (JP 1-02. Source: JP 3-0)

operational reach. The distance and duration across which a unit can successfully employ military capabilities. (JP 1-02. Source: JP 3-0)

personnel services. Those sustainment functions provided to personnel. Personnel services complement logistics by planning for and coordinating efforts that provide and sustain personnel so that the joint force commander may be optimally prepared to accomplish the mission. (NWP 3-32)

priority intelligence requirement (PIR). An intelligence requirement, stated as a priority for intelligence support, that the commander and staff need to understand the adversary or the operational environment. (JP 1-02. Source: JP 5-0)

protection: Preservation of the effectiveness and survivability of mission-related military and nonmilitary personnel, equipment, facilities, information, and infrastructure deployed or located within or outside the boundaries of a given operational area. (JP 1-02. Source: JP 3-0)

risk tolerance. As specified by the commander, the risk to which friendly forces may be subjected from the effects of adversary action(s); acceptable degrees of risk under differing tactical conditions are high, moderate, and negligible. (NWP 3-32)

sea denial. Prevention of the use of the sea by the enemy. Implies sufficient force is not available to ensure the use by one's own forces, but force is available to deny use to the enemy. (NWP 5-01)

situational awareness. Knowledge and understanding of the current situation which promotes timely, relevant, and accurate assessment of friendly, enemy, and other operations within the battlespace in order to facilitate decision making. An informational perspective and skill that foster an ability to determine quickly the context and relevance of events that are unfolding. (NTTP 3-02.3M)

strategic level of war. The level of war at which a nation, often as a member of a group of nations, determines national or multinational (alliance or coalition) strategic security objectives and guidance, and develops and uses national resources to achieve these objectives. Activities at this level establish national and multinational military objectives; sequence initiatives; define limits and assess risks for the use of military and other instruments of national power; develop global plans or theater war plans to achieve those objectives; and provide military forces and other capabilities in accordance with strategic plans. (JP 1-02. Source: JP 3-0)

support. 1. The action of a force that aids, protects, complements, or sustains another force in accordance with a directive requiring such action. 2. A unit that helps another unit in battle. 3. An element of a command that assists, protects, or supplies other forces in combat. (JP 1-02. Source: JP 1)

supported commander. In the context of a support command relationship, the commander who receives assistance from another commander's force or capabilities, and who is responsible for ensuring that the supporting commander understands the assistance required. (JP 1-02. Source: JP 3-0)

supporting commander. In the context of a support command relationship, the commander who aids, protects, complements, or sustains another commander's force, and who is responsible for providing the assistance required by the supported commander. (JP 1-02. Source: JP 3-0)

surface strike group (SSG). The combining of Navy, and perhaps other maritime capabilities that provides combat effectiveness by providing fire support to allies and joint forces ashore in support to crisis response missions or sustained missions and may be employed in limited non-permissive environments characterized by multiple threats. The baseline organization consists of three surface ships based on mission and availability of assets. (OPNAVINST 3501.316A)

tactical command (TACOM). It is a command authority granted to an allied/multinational maritime commander by an allied/multinational maritime commander with either OPCOM or OPCON. TACOM is authority delegated to an allied/multinational commander to assign subordinate forces for the accomplishment of the mission assigned by higher authority. It involves the responsibility for the conduct of the tasks pertaining to the mission(s), i.e., issuing detailed orders and ensuring their correct execution. It also involves responsibility for the general safety of attached units, although ultimate responsibility remains with the commander officers. Tactical command of units temporarily attached does not include the authority to give them tasks inconsistent with the mission previously allocated to them. (ATP 1/MTP 1)

tactical control (TACON). Command authority over assigned or attached forces or commands, or military capability or forces made available for tasking, that is limited to the detailed direction and control of movements or maneuvers within the operational area necessary to accomplish missions or tasks assigned. Tactical control is inherent in operational control. Tactical control may be delegated to, and exercised at any level at or below the level of combatant command. Tactical control provides sufficient authority for controlling and directing the application of force or tactical use of combat support assets within the assigned mission or task. (JP 1-02, Source: JP 1)

tactical level of war. The level of war at which battles and engagements are planned and executed to achieve military objectives assigned to tactical units or task forces. Activities at this level focus on the ordered arrangement and maneuver of combat elements in relation to each other and to the enemy to achieve combat objectives. (JP 1-02. Source: JP 3-0)

task element (TE). A component of a naval task unit organized by the commander of a task unit or higher authority. (JP 1-02. Source: N/A)

task force (TF). A temporary grouping of units, under one commander, formed for the purpose of carrying out a specific operation or mission. (JP 1-02. Source: N/A)

task group (TG). A component of a naval task force organized by the commander of a task force or higher authority. (JP 1-02. Source: N/A)

task unit (TU). A component of a naval task group organized by the commander of a task group or higher authority. (JP 1-02. Source: N/A)

theater of operations (TO). An operational area defined by the geographic combatant commander for the conduct or support of specific military operations. Multiple theaters of operations normally will be geographically separate and focused on different missions. Theaters of operations are usually of significant size, allowing for operations in depth and over extended periods of time. (JP 1-02. Source: JP 3-0)

time horizon. The specified period established by the commander to focus current operations, future operations, and future plans activities for a given operational task. (NWP 3-32)

unified action. The synchronization, coordination, and/or integration of the activities of governmental and nongovernmental entities with military operations to achieve unity of effort. (JP 1-02. Source: N/A)

Universal Joint Task List (UJTL). A menu of capabilities (mission-derived tasks with associated conditions and standards, i.e., the tools) that may be selected by a joint force commander to accomplish the assigned mission. Once identified as essential to mission accomplishment, the tasks are reflected within the command joint mission essential task list. (JP 1-02. Source: JP 3-33)

Acronyms and Abbreviations

AADC area air defense commander

ACA airspace control authority

ACE aviation combat element

ACP allied communication publication

AD air defense

ADCON administrative control

AO area of operations

AOR area of responsibility

ASW antisubmarine warfare

B2C2WG boards, bureaus, centers, cells, and working groups

C/JFMCC combined/joint force maritime component commander

C/JTF combined/joint task force

C2 command and control

CACO casualty assistance calls officer

CAP crisis action planning

CATF commander, amphibious task force

CBRNE chemical, biological, radiological, nuclear, and high-yield explosives

CC critical capability

CCDR combatant commander

CCIR commander's critical information requirement

CE command element

CENTRIXS Combined Enterprise Regional Information Exchange System

CF critical factor

CJCS Chairman of the Joint Chiefs of Staff

NWP 3-32

OCT 2008 LOAA-2

CJTF commander, joint task force

CLF commander, landing force

CMC Commandant of the Marine Corps

CMO civil-military operations

CNO Chief of Naval Operations

COA course of action

COCOM combatant command (command authority)

COG center of gravity

COMNAVFOR commander, Navy forces

CONOPS concept of operations

CONPLAN concept plan

COP common operational picture

COS chief of staff

CR critical requirement

CS civil support

CSG carrier strike group

CTF commander, task force

CV critical vulnerability

CWC composite warfare commander

D3A decide, detect, deliver, and assess

DCOS deputy chief of staff

DIME diplomatic, information, military, and economic

DIRLAUTH direct liaison authorized

DOD Department of Defense

DON Department of the Navy

DP decisive point

EEZ exclusive economic zone

ESG expeditionary strike group

NWP 3-32

 LOAA-3 OCT 2008

FFIR friendly force information requirement

FHA foreign humanitarian assistance

FHP force health protection

FRAGORD fragmentary order

GCC geographic combatant commander

GCE ground combat element

GFMIG Global Force Management Implementation Guidance

HNS host-nation support

HSS health service support

IGO intergovernmental organization

IM information management

IO information operations

IPOE intelligence preparation of the operational environment

ISR intelligence, surveillance, and reconnaissance

JDEIS Joint Doctrine, Education, and Training Electronic Information System

JFACC joint force air component commander

JFC joint force commander

JFLCC joint force land component commander

JFMCC joint force maritime component commander

JFSOCC joint force special operations component commander

JNCC joint network operations control center

JOA joint operations area

JOPES Joint Operation Planning and Execution System

JP joint publication

JTF joint task force

JWICS Joint Worldwide Intelligence Communications System

KMO knowledge management officer

LCE logistics combat element

NWP 3-32

OCT 2008 LOAA-4

LOC line of communications

LOO line of operations

MAGTF Marine air-ground task force

MARFOR Marine forces

MARFORCOM Marine Corps Forces Command

MARFORPAC Marine Forces, Pacific

MCC Marine component commander

MCDP Marine Corps doctrinal publication

MEB Marine expeditionary brigade

MEF Marine expeditionary force

MEU Marine expeditionary unit

MHQ maritime headquarters

MOC maritime operations center

MOE measure of effectiveness

MOP measure of performance

MWR morale, welfare, and recreation

NAVFOR Navy forces

NAVNETWARCOM Naval Network Warfare Command

NCC Navy component commander

NEO noncombatant evacuation operation

NGO nongovernmental organization

NIPRNET Nonsecure Internet Protocol Router Network

NLC Navy logistics command; Navy logistics coordinator

NPP Navy planning process

NSFS naval surface fire support

NWP Navy warfare publication

OA operational area

OGA other government agency

NWP 3-32

LOAA-5 OCT 2008

OLC operational level of command

OPCOM operational command

OPCON operational control

OPGEN operation general matter

OPLAN operation plan

OPORD operation order

OPSEC operations security
OPTASK operation task
OTC officer in tactical command
PA personnel accountability
PAO public affairs officer
PCR personnel casualty report
PIR priority intelligence requirement
PMESII political, military, economic, social, information, and infrastructure
PO peace operations
POLAD political advisor
PR personnel recovery
RC reserve component
RFI request for information
ROE rules of engagement
RUF rules for the use of force
SA situational awareness
SAP special access program
SCC Service component commander
SecDef Secretary of Defense
SECSTATE Secretary of State
SIPRNET Secret Internet Protocol Router Network
SJA staff judge advocate
NWP 3-32
OCT 2008 LOAA-6
SLOC sea line of communications
SNDL standard Navy distribution list
SOP standard operating procedure
SORM standard organization and regulation manual
SPMAGTF special purpose Marine air-ground task force
SPOD seaport of debarkation
SSG surface strike group
STW strike warfare
SUPSIT support situation
SUW surface warfare

TACOM tactical command
TACON tactical control
TE task element
TF task force
TG task group
TPFDD time-phased force and deployment data
TSCP theater security cooperation plan
TST time-sensitive target
TTP tactics, techniques, and procedures
TU task unit
UCP Unified Command Plan
UJTL Universal Joint Task List
UN United Nations
UNCLOS United Nations Convention on the Law of the Sea
US United States
USC United States Code
USFF United States Fleet Forces
USJFCOM United States Joint Forces Command
NWP 3-32
LOAA-7
USPACOM United States Pacific Command
USSOCOM United States Special Operations Command
WARNORD warning order

Index